T0177137

Understanding Animal Behaviour

What to Measure and Why

All students and researchers of behaviour – from those observing animals behaving freely in the field to those conducting more controlled laboratory studies – face the problem of deciding what exactly to measure. Without a scientific framework on which to base them, however, such decisions are often unsystematic and inconsistent. Providing a clear and defined starting point for any behavioural study, this is the first book to make available a set of principles for how to study the organisation of behaviour and, in turn, how to use those insights to select what to measure. The authors provide enough theory to allow readers to understand the derivation of the principles, and draw on numerous examples to demonstrate clearly how the principles can be applied. By providing a systematic framework for selecting what behaviour to measure, the book lays the foundations for a more scientific approach for the study of behaviour.

Sergio Pellis is a professor in the Department of Neuroscience at the University of Lethbridge, Alberta, Canada. He has been researching animal behaviour and neuroscience since 1976 and has maintained a research laboratory at the University of Lethbridge since 1990. For the past 13 years, he has been a Board of Governors Research Chair, and was awarded the Speaker Gold Medal for Research in 2014.

Vivien Pellis was an adjunct professor and is a research associate in the Department of Neuroscience at the University of Lethbridge, Alberta, Canada. She has been a university researcher in animal behaviour and neuroscience since 1981, investigating a diversity of behaviours and species.

Between them, the authors have published more than 250 scientific journal articles and chapters, along with a book titled *The Playful Brain: Venturing to the Limits of Neuroscience* (Oneworld Publications, 2009).

Understanding Animal Behaviour

What to Measure and Why

Sergio Pellis
University of Lethbridge

Vivien Pellis
University of Lethbridge

CAMBRIDGE
UNIVERSITY PRESS

CAMBRIDGE
UNIVERSITY PRESS

University Printing House, Cambridge CB2 8BS, United Kingdom

One Liberty Plaza, 20th Floor, New York, NY 10006, USA

477 Williamstown Road, Port Melbourne, VIC 3207, Australia

314–321, 3rd Floor, Plot 3, Splendor Forum, Jasola District Centre,
New Delhi – 110025, India

79 Anson Road, #06-04/06, Singapore 079906

Cambridge University Press is part of the University of Cambridge.

It furthers the University's mission by disseminating knowledge in the pursuit of
education, learning, and research at the highest international levels of excellence.

www.cambridge.org
Information on this title: www.cambridge.org/9781108483452
DOI: 10.1017/9781108650151

First published 2021

Printed in the United Kingdom by TJ Books Limited, Padstow Cornwall

A catalogue record for this publication is available from the British Library.

Library of Congress Cataloging-in-Publication Data
Names: Pellis, Sergio, author. | Pellis, Vivien, author.
Title: Understanding animal behaviour : what to measure and why / Sergio Pellis,
    University of Lethbridge, Alberta, Vivien Pellis, University of Lethbridge, Alberta.
Description: Cambridge, UK ; New York, NY : Cambridge University Press, 2021. |
    Includes bibliographical references and index.
Identifiers: LCCN 2021003174 (print) | LCCN 2021003175 (ebook) | ISBN
    9781108483452 (hardback) | ISBN 9781108705103 (paperback) | ISBN
    9781108650151 (epub)
Subjects: LCSH: Animal behavior–Research.
Classification: LCC QL751 .P454 2021 (print) | LCC QL751 (ebook) | DDC 591.5–dc23
LC record available at https://lccn.loc.gov/2021003174
LC ebook record available at https://lccn.loc.gov/2021003175

ISBN 978-1-108-48345-2 Hardback
ISBN 978-1-108-70510-3 Paperback

Contents

Preface

This is the book we would like to have had available in the 1970s when we first started studying the behaviour of animals. When conducting one of the first detailed studies of play behaviour in a bird, the Australian magpie, we found precious little guidance in the literature as to what aspects of behaviour to measure. For more global measurements – such as how much of the day is spent playing – the advice reflected in papers and books that were available at the time, and which still have validity (e.g., Altmann, 1974; Dawkins, 2007; Lehner, 1996; Martin & Bateson, 2007), proved very useful (Pellis, 1981a). More difficult was finding advice on what to measure at a micro level, when the birds were playfully wrestling with one another. We explored a variety of statistical approaches at the time – such as those that assessed the probability that one action followed another (e.g., Delius, 1969) or those that assessed how particular actions clustered together geometrically (e.g., Miller, 1975) – but in trying to apply them, we came up against the persistent problem of how to parse the stream of ongoing behaviour into heuristically meaningful units. We found help from methods that made the selection of such units the product of analysis rather than the starting points of analysis (Golani, 1976), and gained considerable insight into the organisation of complex, multi-component behaviour, such as the playful wrestling of magpies (Pellis, 1981b).

What started with Australian magpies expanded into a life-long journey of studying the behaviour of a variety of species, and in each case, it became clear that Ilan Golani was correct – that the most convincing units of behaviour suitable for measurement resulted from the analysis of how that behaviour was organised. With studying diverse species engaging in a variety of behaviours, enduring principles of how behaviour is organised became apparent. We do not take credit for those principles, as most were articulated before we entered the scene, but our own experiences revealed to us that some of those principles are widely applicable. In this book, we make those principles explicit and show how they can be used to gain insight into how behaviour is organised and, in turn, by gaining deeper insight into the organisation of behaviour, to characterise behavioural units or markers of value for quantitative scoring.

One of the most influential papers in the field of Animal Behaviour was by Niko Tinbergen[1] (1963). Tinbergen advocated that a behaviour performed by an animal can be explained in four different ways: by understanding how it promotes the animal's survival and reproduction (function); how the species to which that individual belongs came to acquire that behaviour (phylogeny); the psychological, physiological, and neural processes that make it possible for an individual to express that behaviour (mechanism); and how those mechanisms emerge over the individual's lifetime (ontogeny). In various forms, this 'four whys' approach has become de rigueur for the field (e.g., Alcock, 2013; Bolhuis & Giraldeau, 2005). Unfortunately, another key component of the argument made by Tinbergen in that landmark paper has fallen by the wayside. Tinbergen argued that description is a necessary precursor to answering any of the four whys (Hinde, 1982). Indeed, Konrad Lorenz,[1] another one of the founders of the discipline of animal behaviour, wrote a paper despairing about the demise of description (Lorenz, 1973). We agree with the view that description is instrumental to the study of the behaviour of animals. However, we have also learned that there are two important considerations to take into account when integrating description into current studies of behaviour.

First, while we agree that knowing what needs to be explained must precede attempts to explain the phenomenon, we have also come to understand that description is an iterative process. Experimentation and further comparative analyses can yield novel insights into the organisation of the behaviour that add to or alter the original description. In this way, descriptions – although necessary starting points – are not immutable, but rather, are works in progress. Second, to begin the descriptive process, organisational principles that have proven helpful in other behavioural analyses may be useful. From such a description, behavioural markers can be abstracted that can then be scored quantitatively. Importantly, this

---

[1] In 1973, the Nobel Prize in Physiology or Medicine was awarded to Nikolaas Tinbergen, Konrad Lorenz and Karl von Frisch. The award was given for their work in elucidating the organisation of individual and social behaviour patterns. Basically, they were instrumental in laying the foundations for the biological study of behaviour (Burkhart, 2005).

principled approach provides an objective framework with which to test the usefulness of the abstracted behavioural markers, not only by the researchers proposing those markers, but also by others.

Most particularly, we aim to show that behavioural markers that are a close reflection of the underlying organisation of the behaviour being investigated are more likely to be useful for experiments and comparisons that require numerical assessments. Consequently, this book is intended not only for novices making their first attempts to describe and measure behaviour, but also for experienced researchers who are switching from studying one type of behaviour to another. To make the process of description and the selection of behavioural markers as explicit as possible, we provide a detailed presentation of the principles that we have found most useful in this endeavour.

There are many researchers who have conducted their studies, whether explicitly or implicitly, with some or all of these principles as their guide (for many fine examples, see *Behavioural Brain Research*, Volume 231, Issue 2, 2012), and wherever possible, we use such examples to illustrate the points we are making in the book. However, we give readers fair warning that the majority of the examples we use are from our own studies. This is simply because we know these examples the best and so can use them to the greatest effect to ensure that we convey the underlying methodological issue to the reader. Also, since this is a book intended to provide a methodological guide for abstracting behavioural markers that can then be quantified, we feel no compunction to provide an exhaustive bibliography. What we do provide are sufficient citations so that readers have a starting point from which to explore the primary literature for themselves.

There is one last point we wish to make clear. In articulating the principles and their uses, we only provide sufficient theoretical detail to make them understandable to readers. To develop each of these principles fully, both with regard to the historical conceptualisations that were involved in their development and the biological underpinnings for their evolution and usage by animals, a different book would be needed. For readers interested in exploring the historical origins of some of these principles and how they are used, we recommend the books by Gallistel (1980), Glimcher (2003) and Mook (1996). Although these sources are dated and the principles discussed need to be integrated with the knowledge gained in the intervening years (e.g., Gomez-Marin & Ghazanfar,

2019), they are still a very good starting point to get the feel for those conceptualisations. However, for our current purposes, it is sufficient that readers be able to understand the basic principles and then judge their value themselves by looking at the behaviour of animals. Does the behaviour make more sense now than it did before? Whether the answer is yes or no, by going through the exercise, the observer will have gained a deeper insight into the behaviour in question. What we provide as a guide is intended as an opening gambit for making explicit some of the principles that underlie the organisation of behaviour. As we show in Chapter 6, there may be additional principles that in the future need to be integrated with the ones that we present in the pages that follow. By making behavioural description and the extraction of behavioural markers as explicit as possible, the framework outlined in this book can be built on as new insights and methods arise in the future.

Acknowledgements

This book has been percolating in the back of our minds for many years as we have come to understand the complexities of how to extract useful behaviour heuristically to measure quantitatively, and how to teach our students to appreciate the difficulties. It was Megan Keirnan from Cambridge University Press who saw the value in pursuing such a project and encouraged us to prepare a proposal. Interestingly, the proposal we submitted was theory heavy and the reviewers suggested that we would likely reach a broader audience by making the text more practical with real-life examples of animal behaviour to illustrate the conceptual issues we want to convey. They were right, and this volume attempts to follow the spirit of that advice. Certainly, this approach makes the book more readily accessible to senior undergraduate and graduate students. We thank Megan, the reviewers and our students for their valuable feedback and advice.

Many skilled observers of animal behaviour have influenced our thinking and approach to studying behaviour. Some have been mentors and collaborators, such as Philip Teitelbaum and Ian Whishaw; some have been instrumental in training us, most particularly, Ilan Golani; and some have influenced us via their research papers and other writings, such as Robert and Caroline Blanchard, Valerius Geist and Bill Powers, who we were lucky enough to get to know in person; and others who preceded our active research careers, such as Jakob von Uexküll and Niko Tinbergen. There are many others who have been important influences, but we simply cannot name them all for fear of leaving someone out, but as you read the book, the references cited will reveal these folks. We thank them all.

Once we wrote a draft that we were happy to have others read, some colleagues volunteered to read and critique the tome. The valuable corrections, comments, insights and references provided by Heather Bell, Gordon Burghardt, Evelyn Field and J.-B. Leca have been invaluable in improving the final version of the book. In addition, as the book was written with people at the beginning of a career in studying animal behaviour in mind, we tested its readability and comprehensibility on our current graduate students Candace Burke, Jackson Ham, and Rachel Stark, and in a fourth year seminar course comprising undergraduates majoring in animal behaviour or neuroscience. The feedback was favourable; the key message got through and they enjoyed the read. Candace and Rachel were also instrumental in either adapting and modifying existing figures or creating new ones. We thank them all for helping make this book possible. Finally, we want to thank Megan Keirnan, Aleksandra Serocka, Niranjana Harikrishnan and the other members of Cambridge University Press team who have helped in the completion of the book.

1   What Is the Problem and What Is the Solution?

Count what is countable, measure what is measurable and what is not measurable, make measurable.

Attributed to Galileo Galilei (Finkelstein, 1982, p. 25)

From its inception in the 1500s, a key element in the success of the scientific revolution has been measurement. Rendering the intangibles of nature into numerical values has allowed for precise comparisons. The coupling of accurate measurements with the statistical methods that were developed in the late nineteenth and early twentieth centuries led to the ability to test hypotheses, objectively, about the causal processes that underlie observable phenomena. In this regard, the scientific study of the behaviour of animals is no different to any other branch of modern science as perusal of any journals that involve studying animal behaviour attests (e.g., *Animal Behaviour*, *Behavioral Neuroscience*, *Behaviour*, *Behavioural Brain Research*, *Ethology*, and *Journal of Comparative Psychology*).

However, perusal of papers on topics related to animal behaviour that were published in the 1940s, 1950s, and 1960s show that, even though they became increasingly quantitative, they were highly descriptive. In contrast, papers since the 1970s have become increasingly focused on hypothesis testing – organised so as to answer why animals engage in behaviour X. The concerns that were raised by Tinbergen (1963) and Lorenz (1973) on the need for, and demise of, behavioural description have been largely ignored. This has led to many papers in the modern era providing brief and relatively arbitrary definitions of the behavioural markers to be scored for the quantitative testing of the hypothesis being proposed. The core of the rationale provided in most studies concerns the hypothesis being tested and the sampling methods and statistics used. In this book, we make the case that developing and using behavioural markers is itself a hypothesis – a hypothesis that the chosen measures are appropriate reflections of the behavioural phenomenon being studied. Therefore, rather than being the accepted starting points of a study, the chosen behavioural markers should be subject to empirical testing, just like any other hypothesis. For several reasons, that is not routinely the case.

Amassing and analysing more data Over the past 30 years or so, modern digital and computer technologies have revolutionised not only how data are collected, but also how much can be collected and how they can be analysed. For example, for field-based studies, using handheld devices that tap into the global positioning system have been a boon in accurately tracking the movement of animals and their inter-individual spatial relationships (e.g., Tomkiewicz, Fuller, Kie & Bates, 2010). Digital video and audio recordings have become cheap and easy to use in both field and laboratory contexts. Computer-based analysis systems (e.g., *The Observer* from Noldus, *RavenPro* from Cornell University) have now been refined to the point that large quantities of data can be collected in real time. Furthermore, new computational methods for analysing large amounts of quantitative data derived from traditional statistical methods, or the more recent Bayesian approaches, have been developed (e.g., Casarrubea et al., 2018; Garamszegi et al., 2009; Kline, 2013). Combining large data sets with new computational techniques can lead to novel insights, hitherto unreachable (e.g., Anderson & Perona, 2014; Brown & de Bivort, 2018).

One unfortunate consequence of this trend, however, has been to tolerate poorer quality data, since one can always add another factor in a linear mixed model to rule out statistically the influence of some presumed confound. This is not necessarily a bad thing, particularly at the early exploratory phases of a study, when patterns of association are being sought for identifying material worthy of more detailed study – an approach to which we are not averse (e.g., Burke, Kisko, Euston & Pellis, 2018; Stark et al., 2020). Where this becomes a problem is when some broad statistical pattern becomes confused with real understanding of the biological organisation of the system.

Confounding levels of behavioural organisation Irrespective of the quantity of data collected, it needs to be borne in mind that how data are collected greatly influences how those data can be analysed and what questions can be answered (Gomez-Marin et al., 2014; Leonelli, 2019). A good example of how measuring regimes need to be tailored to the behavioural question of interest concerns the duration of the behavioural events to be measured (Altmann, 1974). If you need to know how much time during the day animals are engaged in different activities, the duration of the bouts of the different activities greatly influences how they are best sampled. For example, to estimate the amount of time a goose spends foraging relative to scanning for predators, that is, whether it has its head down cropping grass or has its head up directing its gaze to the horizon can be recorded at some set interval (e.g., every 10 minutes). That is, at the onset of the time interval, you look at the animal and score whether it is grazing or scanning; then at the beginning of the next interval, the scoring procedure is repeated. Once sampling for the day, say from dawn to dusk, is completed, the number of intervals containing grazing and scanning can be tallied and so the proportion of the samples devoted to each activity can be calculated. Such sampling provides an estimate of how much of the day the animals spend in these activities (Pellis & Pellis, 1982). But such instantaneous scan sampling only works well for behaviours like grazing and scanning, bouts of which last for many seconds or even minutes.

For behaviour patterns like the same goose scratching its head with its hind foot or engaging in a brief courtship encounter, such a sampling technique is inadequate. Given the low frequency of occurrence (which may happen from a couple of times to a few dozen times a day) and short duration (from just a few seconds to less than a second), the chances that the behaviour is caught in the snapshot of an instant when it is the time to sample is highly unlikely. Thus, for rare and short-duration behaviours, a more suitable approach is to sample, continuously, throughout the day, recording them whenever they occur (Pellis, 1982). There are well-established guidelines for taking such factors into account for developing scoring schemes that can appropriately sample different kinds of behaviours (see Dawkins, 2007; Martin & Bateson, 2007).

Whether of short duration or long duration, what all these behaviours have in common is that they are mutually exclusive. From a practical point of view, an observer is unlikely to mistake scratching for grazing or grazing for scanning. Moreover, such behaviours have undeniable biological relevance; eating, avoiding being eaten and grooming are all essential for maintaining life. Also, because these behaviours cannot co-occur, there is no ambiguity in scoring them as independent events and in studying their sequential organisation. Engaging in aggression can be similarly fitted into these scoring schemes and assessed for its occurrence and relative juxtaposition with the other behaviours of interest. However, a closer examination of fighting reveals a level of analysis at which the kind of numerical scoring considered above becomes less informative.

Consider a pair of animals, such as two bull elephants, fighting. The sequence of action can be analysed by determining whether behaviour A follows behaviour B more frequently than expected by chance (e.g., Clark & Moore, 1994; Donaldson et al., 2018; Lerwill & Makings, 1971). But how are behaviours A and B abstracted from the stream of behaviour observed? Most of the movements made by the two animals overlap and continually influence one another in a bidirectional manner (Geist, 1978). Yet, despite these empirical problems, most papers simply state 'my definitions of A and B are ...,' give the heuristic criteria for how they were measured and provide no further rationale. The 'behavioural markers' that are selected for quantification are snapshots of what researchers presume reflect the underlying organisation of the behavioural phenomenon being studied. Researchers' biases in selecting behavioural markers in highly dynamic situations such as fighting are likely to have a greater influence than in selecting those from less dynamic contexts, such as reflected in scoring the scanning and grazing of geese. What are likely to be selected are behavioural actions that are readily identifiable and commonly present in the interactions, but these easy-to-score markers may not be a good reflection of the organisation of the behaviour. Many examples will be explored in the pages that follow.

Confusing agreement with biological reality Increasingly, justification for the validity of arbitrarily selected behavioural markers is how robustly they can be recognised and scored repeatedly by the same observer and by independent observers. While intra- and, especially, inter-observer reliability is an important part of characterising useful measures that can be widely used (Burghardt et al., 2012), by itself, it is an insufficient criterion with which to establish whether an abstracted behavioural marker is a valid description of the behavioural phenomenon in question.

Martin and Bateson (2007) use shooting at a bull's-eye to help conceptualise the reliability of scoring behavioural markers between observers and within the same observer. This is also a helpful metaphor with which to think about the quality of the behavioural marker being scored. The close clustering of bullet holes in Figure 1.1a would represent high inter-observer

(a)   (b)   (c)

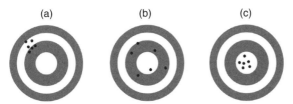

Figure 1.1 A bull's-eye is shown as a representation of how close a
behavioural marker is to the underlying organisation of the
behaviour with the centre being the closest. Inter-observer (or within
observer) consistency is reflected in how close the bullet holes are
clustered together. (a) Precision (highly reliable scoring, but in
this case, off the target). (b) Accuracy (closer to the target, but in this
case, with poor reliability). (c) Precision and accuracy (close to the
target and highly reliable).

reliability compared to the looser clustering in Figure 1.1b. But the clustering in Figure 1.1a is further from the centre than that in Figure 1.1b. In
terms of shooting a target, Figure 1.1a has higher precision (i.e., less
variation among shots), but lower accuracy (i.e., further from the target)
than Figure 1.1b. Simply relying on measures of inter-observer reliability
would lead researchers to view Figure 1.1a as superior to Figure 1.1b.
However, with regard to the biology of the system being measured,
Figure 1.1b is more informative than Figure 1.1a. Of course, in the best
of all possible worlds, Figure 1.1c, which has both precision and accuracy,
would represent the superior measurement scheme. But more often than
not, in a messy world, the actual choice is between Figure 1.1a and b, and
current standards would favour Figure 1.1a because it has higher inter-
observer reliability.

When the level of behavioural organisation under investigation becomes
more prone to subjective judgements as to what should be measured,
increasing the quantity of data collected or relying on inter-observer reliability are poor criteria for passing judgement on what is measured. What is
critical is that the behavioural markers are selected because the researcher
believes that they reflect something important about how the behavioural
phenomenon is organised. In this regard, selection of what to measure is, in
itself, a hypothesis of the underlying organisation of the behaviour, and as
such, should be amenable to being tested. More often than not, the rationale
for selection is not made explicit; only how to measure what is selected is

explicitly stated. But the most critical question that needs to be answered is how closely does the abstracted 'behaviour pattern' or 'behavioural marker' reflect the organisation of the behavioural phenomenon under study. Most currently available books on methods in the study of animal behaviour tend to focus on providing guidance on *how* to measure behaviour (e.g., Dawkins, 2007; Martin & Bateson, 2007), not on *what* to measure. In this book, we provide a framework to make the selection of behavioural markers explicit and so more readily subject to testing.

Some Lessons from Righting

When an adult rat is laid on its back on a flat surface, it will rotate so that it goes from supine (on its back) to prone (with all four of its paws on the ground), that is, it rights itself. Typically, the rotation to prone is cephalo-caudal, starting with the head and ending with the pelvis (Magnus, 1926). Compared to adults, newly born rats are much slower in gaining the prone position, engaging in many, seemingly irrelevant movements that are not present in adults. A simple behavioural marker for assessing how quickly over the course of development animals can achieve the adult-typical form is to measure the time it takes for them to go from supine to prone. This can be done simply: take a video record of the righting and count the number of frames, starting at the frame at which the animal is released, and ending when all four of its paws contact the ground. The number of frames can then be converted to seconds. A complication is that, because righting at early ages can sometimes be very slow, or can even fail to occur altogether, researchers have often chosen some cut-off, such as ending the trial if the animal, after it has been released, has not righted by 15 or 30 seconds. Irrespective of the exact criterion for ending a righting trial, what such studies show is that, with age, animals are increasingly likely to right, and do so with increasing speed, until the timing is indistinguishable from that of adults (e.g., Almli & Fisher, 1977; Altmann & Sudarshan, 1975; Cowan, 1981; Markus & Petit, 1987). There is a practical advantage to this approach, but it comes with a biological disadvantage.

As we discovered by training students to score righting in rats, naïve observers can be quickly taught to count video frames and the scores from multiple students exhibit high inter-observer reliability. Thus, from a practical perspective, this is a 'good' measure; it is reliable across scorers, can be taught easily, can be scored with high precision, and its simplicity allows large samples of animals to be scored rapidly. However, the cost comes with the assumption of what researchers think the measure reveals about the underlying organisation of righting and how that organisation changes with age. By using the time it takes for the animal to right to prone ('time-to-right') as the cardinal marker for the development of righting, it is explicitly or implicitly assumed that righting is a unitary phenomenon that, with maturation of the animal's sensory and motor skills, improves with age. That is, the measure is based on a particular hypothesis about how righting is organised. The problem is, what if that hypothesis is incorrect?

When we first began working in Philip Teitelbaum's laboratory on animal models of Parkinson's disease, righting became one of the behaviours on which we focused. The reason is simple: people with Parkinson's disease have impaired postural reflexes, including the ability to right themselves (Lakke, 1985). Using rats, our goal was to evaluate the behaviour of animals with damage to the neural circuits that are compromised in Parkinson's disease. Naturally, given the literature of the time, we started by using the time-to-right as the relevant behavioural marker. However, this turned out to be an unsatisfactory approach when applied to adult rats with bilateral electrolytic lesions at the level of the hypothalamus. Such lesions damage the ascending dopaminergic neurons, and the disruption of dopamine input to the basal ganglia, anterior to the hypothalamus, results in immobility and catalepsy, symptoms comparable to those of patients with Parkinson's disease. Initially, after the damage, the rats do not right themselves, but with recovery they begin to do so. The recovery involves a complex array of movements, in which there are shifting patterns in how they are integrated until the animals right normally (Pellis et al., 1989). The complexities in how the animals righted through recovery could not be captured by simply scoring the time it took them to right. Since the atypical patterns of righting present in brain-damaged rats could reflect novel compensatory manoeuvres to overcome the Parkinsonian deficiencies, we could not be sure how they related to normal righting.

Teitelbaum's earlier work showed that, both with regard to movement systems and patterns of ingestion, recovery from lateral hypothalamic damage parallels normally occurring development (Teitelbaum, Cheng, & Rozin, 1969; Teitelbaum, Wolgin, De Ryck & Marin, 1976). We therefore turned our attention to how movements occurring during righting changed in rats during ontogeny (V. Pellis, Pellis & Teitelbaum, 1991). Starting with scoring the time the animal took to right to prone, our data concurred with that in the literature that, on average, the speed of righting increased with age. However, the range was extraordinarily large. From the day of birth, infant rats could sometimes right almost as fast as adults. On other occasions, they made repeated movements with their limbs and torso, but failed to gain the prone position. In other cases, they did manage to right to prone, but this took longer than is typical for an adult. The hypothesis that righting involves a unitary pattern of cephalocaudal rotation that improves with age due to the maturation of sensory and motor capabilities did not account for why a newborn could, on occasion, right as fast as an adult! How is that possible if improved righting merely reflects changes in sensorimotor skills?

Working mostly with adult cats, Magnus (1924, 1926) showed that, when falling supine in the air, righting is triggered by the vestibular system (the sensory organ for balance located in the inner ear), or, in the absence of vestibular signals, by vision. When righting on the ground, in addition to those two forms of righting, tactile contact on the upper body triggers righting by the forequarters and contact on the lower body triggers righting by the hindquarters. Since Magnus, another form of tactile righting has been described that involves the trigeminal nerve (a cranial nerve that projects over the face) (Troiani, Petrosini & Passani, 1981). Of the righting systems known to Magnus, he showed that in adults there is a hierarchy. When righting on the ground, irrespective of other sensory signals, vestibular ones have priority access to righting movements. If vestibular signals and vision are blocked, then tactile signals preferentially trigger forequarter righting, with tactile-induced hindquarter righting only occurring if forequarter righting is prevented.

The anomalies in using the time-to-right measure could have arisen from a complex interweaving of these different righting modalities with age. Therefore, following the pioneering work of Magnus (1924, 1926) and others (e.g., Tilney, 1933; Windle & Fish, 1932), we devised ways of testing the capability of each sensory system in triggering righting independently of the influence of other systems in the young of small mammals, including rats (Pellis, Pellis & Nelson, 1992; V. Pellis, Pellis & Teitelbaum, 1991). When doing so, it became apparent that some unique patterns of movement are utilised by each righting system. From birth to weaning, three aspects of righting change until the fully adult pattern is present. First, there is an order of emergence of the different righting systems. Unlike cats, for rats and some of the other small mammals that we studied, vision is not capable of triggering righting at any age. In rats, while all other forms of righting are present from birth, they differ in how closely they resemble the fully adult form. Only trigeminal righting has the typically adult pattern from its first appearance. Second, at earlier ages, unlike fully mature righting, the hierarchy among the different righting systems is incomplete, leading to a simultaneous co-activation of multiple types of righting. Third, vestibular and body tactile forms of righting systems undergo a set of characteristic changes in the combination of head, body, and limb movements that are used over the first three weeks of development until their adult typical forms are consolidated.

Trigeminal righting involves cephalocaudal rotation of the body axis, starting with the head and neck, followed by the shoulders and, finally, the pelvis; this is the same order in infants as it is for adults. Furthermore, this form of righting is completed as successfully in infants as it is in adults. For successful trigeminal righting, the lower part of the infant's snout must maintain contact with the ground, so that its body rotates around that point of contact (Figure 1.2). However, the co-activation of other righting systems and the movements used in the early stages of development can produce actions in the pup that interfere with the successful completion of trigeminal righting. For example, at the onset of tactile-induced forequarter righting, the rat pup's forepaws reach for the ground and pull their forequarters to prone. The immaturity of the pup's forelimbs can lead to a failure to right, so that before the lower side of its face contacts the ground, its forelimbs may lose their grip with the ground and extend upward, away from the ground. This forelimb movement can rotate the pup's shoulders away from the ground and so pull its face away from the ground, interfering with the trigeminal input needed to complete the trigeminal form of righting.

The difficulties are compounded if tactile-induced hindquarter righting is simultaneously triggered with forequarter righting. As in the tactile-triggered forequarter righting that occurs early in development, tactile-triggered hindquarter righting involves reaching and pulling actions by the

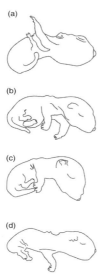

Figure 1.2 A sequence of drawings shows a young marsupial carnivore (the Northern quoll) engaging in trigeminal righting at 40 days of age when righting behaviours begin to emerge. After making side-to-side movements with its head and grasping movements with its forepaws (a), it makes contact with the ground with the anterior of its snout (b and c). Maintaining snout contact with the ground, the quoll rotates to the prone position (d). Adapted from Pellis, Pellis, and Nelson (1992) with permission (Copyright © 1992 John Wiley & Sons, Inc.)

animal's hind paws. The failure of the rat pup to grasp the ground successfully leads to an upward extension of its hind limbs and thus also contributes to the animal rotating its body away from the side of the ground to which its face is closer. Indeed, the co-activation of tactile-triggered forequarter righting and tactile-triggered hindquarter righting can lead to conflicting back-and-forth rotations of the longitudinal axis of the pup's body that is reminiscent of a 'corkscrew' and can prevent either type of righting from being successful in righting the animal's body to prone (Figure 1.3).

The early onset of vestibular righting makes matters worse for the pup, not better. Initially, the rat pup's head is thrust upwards, away from the ground, not towards the ground (Figure 1.4). This reduces the likelihood of trigeminal contact with the ground and makes a successful purchase on the

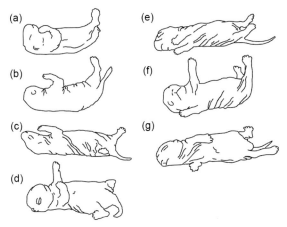

Figure 1.3 A sequence of drawings shows a two-day-old rat pup exhibiting the corkscrew pattern of righting. The pup begins by rotating its forequarters towards the ground with grasping movements of its forepaws. This rotation is accompanied by its righting movements with the hind leg closest to the ground (a and b). However, its hind foot fails to gain traction on the ground and as its leg extends across the midline, it rotates its hindquarters in the opposite direction of its head (c). The pup then rotates its forequarters to the same side as its hindquarters (d), but again, failure by its hind foot to gain traction leads to an extension of its hind leg across its mid-line (e and f). The simultaneous righting movements by the pup's fore- and hind limbs produce a conflicting rotation of its body, which produces the characteristic corkscrew configuration of the body (c and e). In this case, after several failed righting movements such as this, the pup was fatigued and lay inert (g). Adapted from V. Pellis, Pellis, and Teitelbaum (1991) with permission (Copyright © 1991 John Wiley & Sons, Inc.)

ground by the pup's forepaw more difficult. Later in development, both vestibular and tactile-triggered righting by the pup's forequarters switches to the adult-typical form, in which the animal's limbs are tucked into its body and its upper torso rotates around its longitudinal axis. These versions of vestibular and tactile-triggered forequarter righting facilitate, rather than counteract, trigeminal righting. Also, at these later stages, the hierarchy among the righting systems begins to be established, further reducing the likelihood that multiple types of righting are co-activated. As described above, co-activation of multiple righting systems can lead to the animal making movements that are not coordinated with one another,

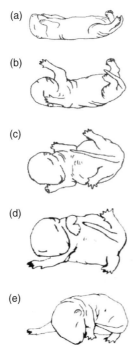

(a)

(b)

(c)

(d)

(e)

Figure 1.4 A sequence of drawings shows a rat pup on the day of birth exhibiting the U form of righting. When released on its back, the pup ventrally flexes its body (a) and then falls to one side (b). Once its forepaw makes contact with the ground (c), the pup rotates its head (d) and body to prone (e). Adapted from V. Pellis, Pellis, and Teitelbaum (1991) with permission (Copyright © 1991 John Wiley & Sons, Inc.)

leading to delayed or failed righting. Given these complex interactions among righting systems, it is no wonder that the time-to-right measurement fails to capture the changes that occur in rats in righting over development or in recovery from the effects of brain damage in adulthood (Pellis, 1996).

Finding Suitable Behavioural Markers to Measure

Thus, although the time-to-right measure is a simple one, and is quick and easy to teach and can achieve a high precision among observers

(Figure 1.1a), as it turns out, it is also one that poorly reflects the organisation of righting and how that organisation changes with age and during recovery from brain damage. As evidenced by the level of training needed to ensure that naïve students can acquire the level of skill needed to recognise and so score complex sequences of righting movements (e.g., Field, Watson & Pellis, 2005; Martens, Whishaw, Miklyaeva & Pellis, 1996), increased accuracy comes with costs in the efficiency of scoring (fewer animals can be scored per unit time) and in the precision of scoring across observers (Figure 1.1b). But how to gain accuracy while increasing precision as much as possible (Figure 1.1c), and, in doing so, have behavioural markers that are not onerous to teach or to use? With regard to the development of righting, this has been achieved in two ways.

First, each type of righting system can be tested separately. For example, to determine the effects of early vestibular sensory experience on the development of the vestibular righting system, the postnatal righting of rat pups was tested in a bath of warm water, in which the sole form of righting possible is that triggered by vestibular stimulation, thus avoiding the confound of triggering multiple righting systems that may be activated when the rat is placed in contact with the ground (V. Pellis, Pellis & Teitelbaum, 1991). Using this more specific testing paradigm, it was shown that, after housing gravid female rats in microgravity (in space flight), their young, when born, were delayed in the onset of vestibular righting, but not in the tactile forms of righting (Ronca & Alberts, 2000). Second, movements that are highly correlated with a particular stage of development can be scored to determine if the rate of maturation is affected by some treatment. For example, in one scheme, body postures typically present at different ages during contact righting (V. Pellis, Pellis & Teitelbaum, 1991) are used to evaluate the rate of maturation of righting. Unidirectional cephalocaudal rotation in which the head is kept away from contact on the ground (and so not involving trigeminal righting) can be used as a marker for the mature, adult-typical form of righting on the ground (Figure 1.5a, axial rotation). An immature stage of righting, in which there is co-activation of forequarter and hindquarter tactile-triggered righting can be readily recognised by the opposing directions of rotation by the fore- and hindquarters (Figure 1.5b, corkscrew). An even earlier appearing immature stage of righting triggered by the vestibular system can be recognised by the initial ventroflexion of the whole body (Figure 1.5c, U-posture). Two studies show

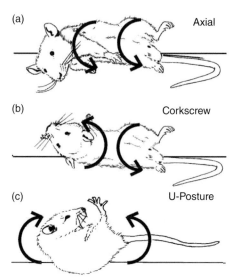

Figure 1.5 Three typical postures associated with different stages of maturation of contact righting are shown. (a) Cephalocaudal axial rotation as is typical of adult vestibular or forequarter tactile-triggered righting. (b) Counterproductive opposing axial rotation as forequarter and hindquarter tactile-triggered righting are co-activated. (c) The initial whole body ventroflexion typical of vestibular-triggered righting typical of the early stages of onset of vestibular righting. Reprinted from Walton et al. (2005) with permission (Copyright © 2005 John Wiley & Sons, Inc.)

how using this simple scoring scheme can reveal the effects of experimental manipulations on the development of righting.

In one study, also involving microgravity, infant rats were exposed to space flight for up to 14 days after birth. Complete, unidirectional cephalocaudal rotation of the body axis reflects the mature form of righting not only of trigeminal righting, but also of vestibular and body tactile forequarter righting, and all these forms are highly likely to occur in their fully mature form during the third week of life (V. Pellis, Pellis & Teitelbaum, 1991). The difference is that trigeminal righting has a relatively mature pattern on first emergence, so unidirectional axial rotation involving snout contact with the ground does not necessarily reflect a mature developmental state (Figure 1.2). However, unidirectional axial rotation in which the head is not in contact with the ground does (Figure 1.5a), and so can be used as a marker for developmental maturity. The rats that were raised in microgravity were less likely to use this pattern than the controls and the difference was permanent in those exposed to microgravity the longest (Walton et al., 2005). Similarly, in the first few days after birth, when rat pups are placed on their backs, vestibular input not only stimulates the head to be thrust upwards, but also the whole body bends ventrally, so that the rat pup adopts a 'U' posture (Figure 1.5c). As the upward vestibular influence wanes, the body tactile forequarter righting becomes predominant, but simultaneously, so does body tactile hindquarter righting. The co-activation of these two forms of body tactile righting leads to the counterproductive back-and-forth movements of the body. This pattern of righting can be readily identified by the corkscrew configuration of the body (Figure 1.5b). As the U posture is more frequent than the corkscrew posture in the first couple of days after birth, the former can be thought of as representing an immature stage of development (V. Pellis, Pellis & Teitelbaum, 1991). Comparing the relative frequency of the U and the corkscrew postures in rats that were born prematurely compared to normal gestational length revealed that the premature rats developed the more mature form of righting (i.e., the corkscrew) more rapidly (Roberto & Brumley, 2014).

With regard to understanding how righting develops, scoring time-to-right yields precision but poor accuracy (Figure 1.1a), and scoring whole sequences of righting yields accuracy but less precision (Figure 1.1b). However, by persisting with the option shown in Figure 1.1b, the sensory and motor patterns that reflect the developmental organisation of righting were identified (V. Pellis, Pellis & Teitelbaum, 1991; Pellis, Pellis & Nelson, 1992). As shown by the examples mentioned, behavioural markers that are relatively easy to learn to identify and score could be abstracted to assess particular aspects of the organisation of righting, such as the relative maturity of animals reared in particular environments. Scoring these measures yields both accuracy and precision (Figure 1.1c). Even though we have used the time-to-right measure as a foil to illustrate how behavioural markers should be selected to reflect the organisation of the behaviour being studied, if this measure is pegged to an appropriate aspect of organisation, it too can yield accurate as well as precise measurements.

For example, in trigeminal righting, a righting system that involves the same sequence of movements across all developmental stages, the time-to-right measure can be useful to assess whether sensorimotor changes with age affect the execution of righting. Video records taken at a higher frame-rate than we used could numerically assess the magnitude of the small age-dependent improvement that we noticed. Similarly, individual variation is another potential factor that can affect performance. For example, Madagascar hissing cockroaches from the same colony were grouped into bold and timid animals based on their responses to various threats. Righting to prone when placed supine on the ground was tested, with the bold animals righting faster than the timid ones. However, this is a composite measure that combines both the latency to begin to right and the time it takes to right once righting is initiated. When these two components of timing were scored separately, it was found that timid cockroaches had a longer delay in beginning to right, but once they began, they right themselves just as quickly as bold animals do (Logue et al., 2009). That is, the timid animals are more inhibited in commencing movements. The same difference in behavioural inhibition has been demonstrated in rodents and has been linked to variation in spontaneously occurring levels of neurotransmitters, such as dopamine (Cools et al., 1990; Pellis & McKenna, 1992). Suppressing such neurotransmitters experimentally delays the onset of initiating righting, but does not affect the speed of righting (Martens, Whishaw, Miklyaeva & Pellis, 1996). The lesson to be learned is that, if the time taken to right to prone is used to measure an appropriate aspect of the underlying behavioural organisation, then it too can be both an accurate and precise measurement.

Some Thoughts on Hypotheses, Hypothesis Testing, and Predictions

A central theme of this book is that proposed behavioural markers are hypotheses about behavioural organisation. However, we have detected some sloppiness in the current literature in how behavioural hypotheses are framed and tested. Therefore, we think it worthwhile to look at what constitutes hypotheses more closely.

In an interesting paper, Hempel and Oppenheim (1948) draw the distinction between an *explanandum*, a Latin term for the phenomenon to be explained, and the *explanans*, the Latin term for the proposed explanation of that phenomenon. The authors use an example to clarify this distinction. One person asks, 'why is there smoke?' and another responds with 'because there is a fire.' In this case, 'smoke' is the *explanandum*, and 'fire' is the *explanans* – the presence of smoke is explained by fire. Any of the behavioural examples used in this book can be viewed from this perspective. For example, in this chapter, we discuss the phenomenon of counterproductive righting movements by the fore and hindquarters in young rats (see Figure 1.3). The question of 'why do young rats perform this corkscrew action?' is the *explanandum*, the phenomenon that needs to be explained. The next phase in the process is to find an *explanans*, an explanation for why the phenomenon occurs.

Before we began to study the development of righting, one explanation for the corkscrew was that, given their sensorimotor immaturity, this was an efficient pattern of righting in infant mammals (Horwich, 1972). That is, the *explanans* is the hypothesis that the corkscrew is an infantile adaptation for righting. In contrast, as described in detail above, our explanation was that due to the immaturity of the central control mechanisms, two forms of righting – one involving tactile-triggered righting by the forequarters and another involving tactile-triggered righting by the hindquarters – are co-activated, leading to counterproductive movements that impede righting. That is, our *explanans* is the hypothesis that the corkscrew is a by-product of immaturity, not an adaptation to overcome the limitations of immaturity (V. Pellis, Pellis & Teitelbaum, 1991). Having two clearly articulated alternative *explanans*, that is, hypotheses, sets the stage for testing their relative explanatory utility. Tests are usually framed in concrete ways by generating predictions that follow logically from the hypotheses.

In this case, if the 'corkscrew as an adaptation' hypothesis were true, then one prediction should be that mammals in which righting is developed *in utero* should not have this pattern of righting. The reason is that if this pattern of righting were an adaptation, only infants that need to right would have this capacity. In contrast, if the 'corkscrew as a by-product of immaturity' hypothesis were true, this would not be the case, as the maturation of the different forms of righting should proceed in a similar manner irrespective of whether the infant lives in an environment that requires it to right or not.

We tested this prediction by examining the development of righting in a small carnivorous marsupial, the Northern quoll (Pellis, Pellis & Nelson,

1992). As a marsupial, quolls are born at an immature stage, comparable to that of a foetus in placental mammals. The newly born quoll then migrates to the pouch area on its mother's ventrum and attaches itself to a teat, where it completes its development until it can move about freely, apart from its mother (Nelson & Gemmell, 2004). At the age at which righting is being developed in foetal quolls, they are firmly attached to one of their mother's teats – skin grows around the edges of their mouths so they cannot be dislodged. Their forepaws are well developed with protruding claws that enable them to hold onto the thick hairs on the mother's ventrum. By removing such infants from their mothers at different ages and testing them on the ground, in the air and in water, as we did with young rats (Pellis & Pellis, 1994; V. Pellis, Pellis & Teitelbaum, 1991), we could evaluate the development of righting in quolls.

Even though when attached to the mother, infant quolls never have an occasion to need to right to prone, when tested, they nevertheless undergo the same pattern of maturation of the different righting systems as rats. Furthermore, in the quolls, when tactile forms of forequarter and hind-quarter righting begin to mature, they are co-activated, resulting in the corkscrew form of righting. This is counter to the prediction made by the 'corkscrew as an adaptation' hypothesis, but is consistent with the 'corkscrew as a by-product' hypothesis. This example highlights two aspects about the relationship between hypotheses and predictions that we think needs to be closely considered when using the hypothesis testing method.

First, both when examining graduate theses and in reviewing papers for scientific journals, we have found that there is often confusion between hypotheses and predictions. A statement such as 'we hypothesise that A should follow B' does not constitute a hypothesis for the simple reason that there is no explanation as to why A should follow B. A hypothesis needs to contain an *explanans*, otherwise the proposed relationship between A and B are simply statements about the relationship between components of the phenomenon being studied, the *explanandum*. This confusion is more likely to be found in descriptive studies that derive behavioural markers for quantitative scoring. What should be in the forefront of the researcher's mind is whether what is being posited contains an explanation for the phenomenon being studied. If so, then the prediction about the relationship between A and B should follow from that explanation. In the example of the corkscrew, because of the different causal processes invoked by the two hypotheses, different predictions can be made about the expected presence of the corkscrew dependent on the environment in which the young develops. Consequently, the causal processes in a hypothesis should explain why A should follow B, and, if so, the statement that A should follow B can be posited as a prediction. Measuring whether A follows B in the real world would then be a test of the hypothesis.

Second, and again, from reviewing many submitted papers, time and again we have noticed that many predictions are weakly pegged to the posited hypothesis. The conclusion is often in the form of 'the predictions are consistent with our hypothesis'. However, as reviewers, our reaction is, 'sure, but they are also consistent with three other hypotheses that we can posit.' The most convincing predictions are ones that are unique to particular hypotheses, so that, if they are supported or contradicted, they provide strong evidence for whether a particular hypothesis has merit or not. If both the 'corkscrew as an adaptation' hypothesis and the 'corkscrew as a by-product' hypothesis predicted that all young animals in all contexts should exhibit the corkscrew, then a positive finding that all young animals perform the corkscrew, even though consistent with both hypotheses, would not be a particularly useful test of either hypothesis. We should note that we are not posing as overly insightful researchers in giving this advice. Often, it is the constructive criticisms from reviewers that lead us to develop more refined hypotheses and mutually exclusive predictions. Finding the flaws in the work of others is often much easier than finding the flaws in one's own work. Let us illustrate with an example.

Spider monkeys shake their heads from side to side so as to facilitate amicable social contact. This occurs frequently during vigorous play fighting, and so it is common during the juvenile period. Occasionally, juvenile spider monkeys use headshakes during non-social locomotor play. The phenomenon needing explanation was the occurrence of these seemingly out-of-context headshakes (the *explanandum*). Based on several prevailing views about why animals sometimes perform out-of-context behaviours during play, we developed three hypotheses (the *explanans*). One hypothesis was derived from the view that play trains animals to deal with unpredictability, another hypothesis was derived from the view that immaturity can result in misdirected signalling, and the final hypothesis was derived from the view that animals may use signals to self-regulate their own emotional state (Pellis & Pellis, 2011). With three competing hypotheses, it was a challenge to develop predictions that would not simply be consistent with more than one hypothesis but predictions that were mutually exclusive, so that, if they turned out in a particular way, they would not only provide confirmatory evidence for one hypothesis but would also disconfirm the other two. It took three rounds of revision with critiques and suggestions from reviewers to finally develop a set of predictions that had this highly discriminative structure (see Table 1.1).

The two hypotheses that we thought would be more likely to be involved – the training for the unexpected and the misdirection due to immaturity – turned out to have the least support; rather, it was the self-regulation hypothesis that received the strongest support. Without strong predictions that provided mutually exclusive tests of the three hypotheses – with some of those predictions suggested to us by diligent reviewers – very different conclusions would likely have emerged from this study. For another insightful example, see Waterman (2010). In the course of the book, we endeavour to use this strong version of behaviour testing to test particular hypothesised behavioural markers.

The Way Forward

It should be noted that the problem of identifying suitable behavioural markers to score is the same whether we are dealing with two or more animals engaging in an interaction or if the body parts of a single animal are combined in various ways. The first principle to digest and use then is that a behavioural marker is a reflection of the hypothesis that the researcher has in mind as to how that marker reflects the underlying organisation of the behaviour to be studied. As such, the hypothesis needs to be made explicit and tested. After all, the marker could be a poor reflection of that organisation, or, alternatively, the hypothesis about the organisation of the behaviour may be wrong or deficient.

This book will explore principles that can guide the researcher in characterising features of the organisation of the behavioural phenomenon being studied and so identify markers that would be suitable for scoring quantitatively. We do not pretend to provide an exhaustive list of principles. Indeed, in the last chapter, we offer clues to new principles that will likely need to be included in the future. Nonetheless, the principles that we

Table 1.1 Three hypotheses and their associated predictions regarding the non-social headshakes (HS) by spider monkeys are detailed. Adapted from Pellis & Pellis (2011) with permission (Copyright © 2011, American Psychological Association)

	Immaturity	Increasing variability	Self-regulation
Hypotheses	HS in a non-social context arise from being misdirected to incorrect targets.	HS in a non-social context are performed to experience unpredictability.	HS in a non-social context are used to motivate action in unfamiliar contexts.
Predictions	**1a.** HS should occur frequently when playfully mouthing both social partners and inanimate objects, especially when very young. **2.** HS in non-social contexts should be absent in adults. **3.** Social and non-social HS should be structurally similar. **4.** No differences predicted about actions followed or not by non-social HS.	**1a.** HS should occur frequently in all contexts, especially when the young are more independent of their mothers. **2.** HS in non-social contexts should be absent in adults. **3.** Social and non-social HS should be structurally similar. **4.** No differences predicted about actions followed or not by non-social HS.	**1a.** HS should occur in a contextually correct manner. **1b.** Non-social HS should mostly occur when young monkeys are making their first forays away from their mothers. **2.** HS in non-social contexts should be present in adults. **3.** Neither similarity nor difference predicted between social and non-social HS. **4.** There should be hesitancy in actions performed in association with non-social HS.

do provide have proven to be valuable guides in our own work and that of many others. Most importantly, their utility has stood the test of time. Once the four main principles are developed over the next three chapters, Chapter 5 will show how they can all be brought together so as to aid in understanding the organisation of a complex behaviour and to identify useful behavioural markers that can be abstracted for quantitative studies. However, in all cases, in applying one or more of these principles, we stress that the resultant markers are hypotheses about the underlying organisation and so need to be tested. In the final chapter, we will also explore some novel methodological approaches for testing such hypotheses.

2 Behaviour as a Means, Not an End

We must trust the evidence of our senses rather than our theories and theories as well, as long as their results agree with what is observed

Aristotle (Lloyd, 1968, pp. 68–79)

A two-millimetre tick has positioned itself in a shrub, and, as a dog walks past, it lets go of its hold on the shrub and falls on the hapless mutt, from whom it obtains a meal of blood. This is a female tick's strategy for provisioning its brood of eggs with appropriate nutrients. If a photograph were taken from the tick's vantage point, it would reveal large green, flat structures (leaves), thick brown cylinders (branches), a bluish tinged backdrop above (sky) and, below, a motley collection of browns, yellows and uneven textures (the ground, which is covered with fallen vegetal detritus). The dog is a large, lumbering, furred quadruped with a nose at one end and a tail at the other. But the tick sees none of this. Indeed, the tick has no eyes, so what does it sense? First, it detects a vertical surface (the shrub), which induces it to climb upwards. Second, it ignores the plethora of possible stimuli, and, by doing so, it can remain attached, unmoving, to the twig. Third, when it detects the odour of butyric acid, the tick releases its grip on the shrub and falls downward. The third step is critical, as butyric acid exudes from the skin of mammals, like the dog, on which the tick lands. Once in contact with the dog, the tick moves to find a patch of bare skin, which is detected by being warmer than the surrounding area covered in fur. The tick then uses its mouthparts to drill into the dog's skin and gain access to its blood. Studies with fluids held in membranes have revealed that it is simply the warmth of the fluid that attracts the tick's sucking, not the nutrients contained in the dog's blood. If the tick misses its mammalian target, it walks about, finds a vertical surface and climbs back upward.

This classic example was used by Jakob von Uexküll to illustrate how different animals 'see' the world (von Uexküll, 1934/2010). To the tick, the world is composed of vertical surfaces to climb, 'bags' that exude butyric acid on which to pounce and warm sacks that contain fluids to suck. All the detail captured by our camera is not part of the world sensed by the tick. What von Uexküll showed, with this example, is that each species has its own *Umwelt*, a view of the world (for recent explorations of von Uexküll and his concept of the *Umwelt*, see Berthoz & Christen, 2009 and Brentari, 2015). From this perspective, when a human observer watches an animal engage in some behaviour of interest, the first problem is to determine to what in its environment the animal is attending. This requires knowledge about the sensory information the animal is able to detect but also which of all the sensory information that the animal can detect it actually processes and uses to guide its behaviour.

Determining what sensory information is available to the species being studied is a physiological problem. As human observers, if that species is using sensory organs that we share, then we may have some intuitive clues of what may be being sensed. However, in some cases, the animal may be using sensory organs that are alien to us, and in such situations, we have no intuitive basis for gaining insight into how the animal may be navigating its world. For example, consider rattlesnakes hunting at night, when their vision may not be adequate to detect and strike at prey. As pit vipers, rattlesnakes have special heat-sensing organs on their faces that function as accessory 'eyes' that enable them to extend their vision into the infrared wavelengths. Thus, in complete darkness, rattlesnakes can 'see' nearby prey by the heat it exudes. In low light conditions, the heat from the prey supplements the visual image from the rattlesnake's eyes, which enables it to gain a better resolution of its location so as to direct its strike more accurately (Clark, 2016).

As both the tick and the rattlesnake use sensory cues that are alien to us, they serve as important case studies for us as observers to evaluate, carefully, the sensory world of species whose behaviour we wish to understand. However, there are also differences between these two species that further illuminate what we need to understand about a specific animal's world. For the tick, a mouse, rabbit, dog or person would be just as good. In fact, the bigger the animal, the greater the odour of butyric acid exuded and so the more likely the tick is to drop, 'lured' by a bigger meal. In contrast, for the rattlesnake, if confronted by a coyote or a human, even though they are bigger and so would exude a larger heat signature, they would not be suitable prey items; a mouse-sized prey would be just about right. That is, while all sources of butyric acid are suitable prey for the tick, not all sources of heat are suitable prey for the rattlesnake. Indeed, for the rattlesnake, a large heat-producing animal could be a threat more likely to trigger it rattling as a warning to that large animal to keep away. This means that knowing the sensory input available to the animal being studied may not be sufficient to explain the animal's behaviour. Rather, the sensory input is filtered to form the *Umwelt* (Berthoz, 2009), and it is that filtered sensory input that forms the animal's perception that guides its behaviour.

For larger species of rattlesnakes, it is small mammals such as mice that are preyed upon, but as juveniles, the same individuals typically prey upon small lizards (Clark, 2016). As ambush predators, rattlesnakes seek out the locations where their prey are likely to congregate and then lie in wait for them. Both adult and juvenile rattlesnakes do this by using chemical cues which are collected and processed with the vomeronasal organ (VNO) (Burghardt, 1980).[1] The snake follows the odour trail left by its prey to its refuge. Here, their odour is in a higher concentration than elsewhere. Thus, while snakes of both ages use smell to track down the refuges used by prey, they focus on different odours, ones that are characteristic of lizards in one case and ones characteristic of mice in the other. That specific sensory inputs create the *Umwelt* is well illustrated by those quintessential olfactory trackers, bloodhounds. Often depicted in classic Hollywood movies, such as *Cool Hand Luke* (1967) and *The Defiant Ones* (1958), the olfactory sensitivity of these dogs is put to good use to track escapees from prisons. But for the dogs to be useful, they need to know the smell of their target. The prisoner's clothing or bedding is given to the dogs to sniff and they then use this olfactory 'image' to track the path taken by that prisoner over terrain that may contain a multitude of different odours. To be successful in their hunt, the dogs need to filter out the majority of the odours that they sense so as to maintain their focus on the prisoner's specific odour. A wide variety of studies with many different species have shown that such sensory filtering is critical for animals to focus on the cues critical to their survival and reproduction (Ewert, 2005). That is, even when the same sensory input is used, to be of value in relating the animal's behaviour to

---

[1] The olfactory system with which we are most familiar links the sensory epithelia in the nostrils to the primary olfactory bulbs. Many vertebrates, such as reptiles and some mammals, have an additional olfactory system, used to detect large, non-volatile molecules. The vomeronasal organ links the sensory epithelia of two pits on the anterior surface of the roof of the mouth to the accessory olfactory bulbs (Cooper & Burghardt, 1990). In snakes, the tines of the forked tongue collect such molecular information by flicking the tongue, then passing the tines over the pits in the mouth and so act as a chemoreceptor (Meredith & Burghardt, 1978).

what it senses, it is necessary to determine what sensory information constitutes the *perceptions* that guide behaviour (Berthoz, 2009).

Identifying the Perceptions Salient to the Animals Being Studied

As noted, what is sensed is a physiological problem, but what is perceived is a problem that requires a behavioural perspective. Of the myriad stimuli available to the animal, what needs to be determined are the ones to which the animal is attending. An important clue can be derived from the movements of the animal in space. For example, as noted earlier, rattlesnakes track down potential prey by following their odour trails (Clarke, 2016). Without knowing that the snake is following an odour trail – which without suitable equipment the human observer is not privy to – the path travelled could be interpreted in multiple ways.

The snake crawls over a hillock of stones, beneath a patch of cactus, across an open stretch of sand and finally into a clump of bushes. It could be engaging in a semi-random search pattern, or, if hunting during the day, the search could be interrupted by periodically moving underneath vegetation so as to limit its continuous exposure to the sun, or the path could represent one that has previously led to hunting success. A combination of tactile, proprioceptive, visual and olfactory cues could provide the snake with the means to create new paths or follow old ones. However, once the researcher knows that the snake is hunting and that for this ambush predator the first phase involves finding the refuge of a prey, its movements begin to make sense. This is especially true once physiological studies have identified that olfaction is the primary means used to find such refuges. The path is clearly not a random search interspersed with protective stops; rather, the snake is following the odour trail left by its prey, and by doing so, replicates the path travelled by it. Even though the snake may be capable of detecting a variety of sensory stimuli from various modalities and, indeed, a variety of odours, like the bloodhound, it ignores all these sensations and focusses its attention on the prey-specific odour. The tracking behaviour is also instructive with regard to the relationship between salient perceptions and behaviour.

Once the salient odour is detected, the snake follows that odour to its source. It does so by collecting olfactory input and moving towards the direction with a higher odour concentration. The scent is collected with both sides of the VNO and then by comparing the relative strength of the scent between the two, the snake can track the path travelled by its prey (Golan et al., 1982). If the scent is stronger on the right side of the VNO, the snake veers to the right, if stronger in the left it veers to the left, with a straight path ensuing when the chemical signal equal on both sides of the head and move towards the stronger signal. Using this simple, negative feedback mechanism, the snake can use the odour left behind by the prey to track it down to its refuge, and since this is where the prey spends most of its time, this location has the highest concentration of its odour. This provides the cue for the snake to stop moving, adopt its coiled ambush posture and wait for its prey to return. Then, when the prey returns, the snake will use visual and/or heat signals to judge the direction and distance of the venom injecting strike.

The negative feedback mechanism that enables the rattlesnake to track down its prey's home turns behavioural analysis on its head. The most common way to think about behaviour is that it represents the output triggered by sensory stimuli, coupled with varying degrees of intervening neural/psychological processing (Alcock, 2013; Kolb & Whishaw, 2015; Staddon, 2016). From this perspective, sensory stimuli control behaviour. For example, in a standard learning paradigm, a rat is trained to learn that when the small light in front of its cage is turned on, a pellet is released in a trough at the other end of the enclosure. Once this association is learned, the rat will reliably go to the trough when the light is switched on. That is, the light stimulus controls the behaviour. The example of the rattlesnake shows the exact opposite relationship between sensory stimuli and behaviour. The side-to-side movements of its head and the path the snake travels are present so as to control its perceptual input – keep the chemical signal equal on both sides of the head and move towards the stronger odour signal. That is, the behaviour controls the perception. In the more commonly used approach, the relationship between perception and behaviour is linear (stimulus $\rightarrow$ behaviour), whereas in the latter, the relationship is bidirectional or circular (stimulus $\longleftrightarrow$ behaviour).

The negative feedback in circular causal systems has been established at the physiological level since the pioneering work of Claude Bernard (1865/ 1927) and is a mainstay of modern-day physiological science (e.g., Sherwood & Ward, 2018). For example, the rate of ventilating our lungs is regulated to maintain a constant, low level of carbon dioxide in our blood (Horrobin, 1970). The application of circular causation to behavioural studies, however, has had a much patchier history (Cziko, 2000;

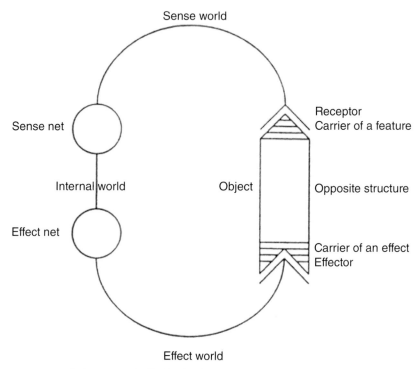

Sense world

Sense net

Internal world

Effect net

Effect world

Object

Receptor
Carrier of a feature

Opposite structure

Carrier of an effect
Effector

Figure 2.1 Jakob von Uexküll's feedback cycle, linking perception and action into one circuit. From von Uexküll (1934/2010)

McFarland, 1971; Turner, 2007). One of the first explicit attempts to integrate the negative feedback relationships between perceptions and actions was by von Uexküll (1934/2010). Indeed, he formalised this circular causality in a diagram (Figure 2.1).

Von Uexküll's diagram shows a circular causal relationship between the animal's perceived world (*Umwelt*) and the behaviour performed or its action world (*Wirkwelt*). The salient perceptions elicit particular behaviours and by executing those behaviours, the perceptions are controlled. The circuitous route travelled by the rattlesnake makes sense once it is understood that its actions arise as attempts by the snake to control the salient olfactory perception. The snake crawls over the hillock of stones and crawls beneath the patch of cactus because that is where the prey animal travelled and left its odour trail. In this conception of the perception–behaviour relationship, behaviour is envisaged as providing compensatory actions that maintain a perception constant, which, in the case of the rattlesnake, involves maintaining an equal concentration of scent between the two sides of the VNO. A useful theory that has formalised this approach is 'Perceptual Control Theory' (PCT) (Powers, 2005). For present purposes, the most useful aspect of this theory is that the organisation of behaviour is hypothesised to arise from the perception(s) that is being controlled. Thus, when studying a new behaviour, the first goal is to identify that perception. This constitutes the controlled variable (CV). The reference level at which that CV is maintained is often very specific and reveals that it is indeed that CV that explains the behaviour. Again, righting provides a good example.

When righting, the animal rotates its body from a supine to prone position (see Figure 1.2) and it could be argued that the inverted position triggers righting and achieving paw contact on the ground triggers cessation of righting. This would be a simple reflex action in which specific sensory stimuli control the behaviour performed (Sherrington, 1906). However, consider dropping a cat supine in the air: after falling for a short distance, it begins to rotate, cephalocaudally. But the cat achieves the prone position before it lands and remains in this orientation until it does land (Magnus, 1926). Therefore, it cannot be that it is paw contact on the ground that inhibits further rotation, although the prone position of the cat's head or the air resistance felt on its ventrum could provide such stop signals. The latter is unlikely because if the cat were thrown upward while inverted, it rotates to prone while ascending, that is, while experiencing air resistance on its ventrum (Magnus, 1924). Putting a blindfold on the cat blocks visual information, ensuring that the righting is guided by vestibular information alone. In this context, could it be the vestibular signal of the head being prone that triggers a reflexive cessation of rotation? There is an alternative possibility.

From the perspective of PCT, the inverted position of the head produces a vestibular error signal that is corrected by the cat rotating its head until it is prone, thus eliminating the error as the head is turned $180°$. That is, the prone position of the head is the CV, the implication being that the amount of rotatory movement by the head is determined by the amount of deviation from the horizontal relative to gravity. That it is the satisfying of such a reference signal emanating from the vestibular apparatus that determines how much movement is performed is illustrated by a marvellous experiment. Rabbits were spun in a drum so that they experienced ten seconds of their head being tilted by $18°$ before being inverted and released. When righting, they rotated their heads $162°$. That is, they gained and maintained a posture that was 18° from horizontal, satisfying the altered vestibular reference signal (Brindley, 1965). It is correcting the error signal that leads to the cessation of rotatory movements, not the achievement of being prone *per se*. The behaviour functions to gain and maintain the value of the reference signal associated with the CV (Powers, 2005). From this conceptual stance, the problem at the beginning of a behavioural analysis is to identify the CV that is being maintained.

The CV can be identified by what is unchanged over the course of the behaviour. In the case of the rattlesnake, what is controlled is the balance of chemical input between the two sides of the VNO. If the input is equal, the error is zero (the concentration on the right minus the concentration on the left), with a perfect linear path maintaining this difference at zero. Oscillation around the preferred value – in this case, zero – for the controlled variable, is inescapable in an imperfect world. How quickly the error can be detected depends on the physiological resolution capabilities of the sensory organs involved and so how quickly the correction can be enacted. For example, for an olfactory system that can only detect differences greater than 1000 ppm (parts per million), differences of 100 ppm would not be corrected, but if the sensory organ can detect differences of 50 ppm, then it would. Depending on the sensitivity of the sensory system, oscillations can be small or large, but in both cases, they would cycle around the preferred value (Figure 2.2). The same applies to the effector organs that produce the behaviour. That is, how quickly they can be activated and how rapidly the actions can be executed will determine how big the drift away from the preferred value will be before corrective action can begin to diminish the error (Golani, 1976).

Irrespective of the noisiness of the system, a key organisational principle of PCT is that identifying the CV can explain much of the overt behaviour observed. This insight, in turn, influences the selection of behavioural markers useful for numerical scoring. To illustrate the use of this principle, we will describe courtship behaviour in the Cape Barren goose, a species of goose that is endemic to the Southeastern shoreline of Australia, especially on the offshore islands. The adults of this species form long-lasting male–female pair bonds and control territories that they use for food (freshly growing grass) and as nesting sites (Dorwood, 1977). Courtship involves the male circling the female, a feature seemingly common to many water birds (Johnsgard, 1965). If male circling of a female is an intrinsic feature

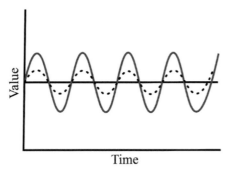

Figure 2.2 The graph shows a variable (Y-axis) changing over time (X-axis). The straight line represents the preferred value for the CV that the animal maintains. Disturbances to the maintenance of the CV lead to movement away from the preferred value, with compensatory action by the animal regaining the preferred value. These ranging oscillations around a preferred CV can be large (solid line) or small (dashed line), depending on how effectively the animal can counteract the disturbances.

of the organisation of courtship, then scoring the frequency of circling could be a good behavioural marker for the reproductive state of the birds, with intensity differences in circling across species potentially being a good way of assessing species differences in bonding patterns and so on.

When we first began studying a free-ranging population of this species at a nature reserve (Pellis & Pellis, 1982), we did see courtship and it did involve a male circling a female. However, we also saw males follow females in a straight line, and, in some cases, males tracing zigzag paths as they followed females. Moreover, the relative frequency of straight lines, zigzags, and circles varied across pairs and over the course of the mating season (until the female laid all her eggs). Consulting handbooks on waterfowl behaviour did not shed light on these varied movements of the male around the female. Recalling the old saying that 'it takes two to tango', we suspected that it could be that the correlated actions by both partners needed to be taken into account to make sense of why different paths were traced by the males. Therefore, we filmed sequences of male–female courtship interactions for detailed analysis (Pellis, 1982).

As the movements of the pairmates co-occur and the goal of finding the CV is to identify what changes the least, this creates difficulties for human observers who are biased to detect change, not constancy (Kolb, Whishaw & Teskey, 2019). To this end, we learned how to use a choreographic technique for notating movement in three spatial dimensions and over time (Eshkol & Wachmann, 1958). The movement of the body parts of each partner and the spatiotemporal relationship between the partners are notated on a page resembling a musical score. This descriptive method forces the human observer to record not only what changes over the course of the interaction, but also to notice what remains constant (Golani, 1976). For details on how this notation system works, see Appendix A, but for the present illustration of the principle we will simply show what was revealed.

The male goose approaches the female, lowers his neck and orients his bill towards her rump, but as he gets close to making contact, the female rotates, pivoting her rump away from the male. The male keeps following the female and so circles around her (Figure 2.3). However, if the female begins to walk away from him when he is at a distance of one or two body lengths away, then, as he keeps his beak oriented to her rump, he traces a linear path in space. If she begins to walk away when he is closer, say,

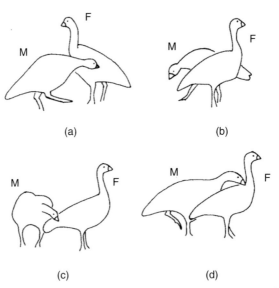

(a)   (b)

(c)   (d)

Figure 2.3 Sequence of drawings shows a male approaching a female, directing his beak towards her tail. (a) As he is closing the distance between his beak and her tail, she rotates 180°, leading him to walk in a circular path (b), until she stops and he contacts her tail feathers (c). Adapted from Pellis (1982) with permission (Copyright © 1982 Cognizant Communication Corporation).

within half a body length away, she may swerve first to the right and then to the left as she walks, making it difficult for him to contact her rump, but, as he follows, he traces a zigzag path. In all three sequences, what is constant is that the male maintains the orientation of his bill towards the female's rump, with the circling, straight, and zigzag paths that he follows all being the consequences of the evasive actions taken by the female that blocks him from contacting her rump.

Therefore, circling by the male goose is not an intrinsic organisational feature of courtship, but rather, it arises because of the compensatory action he takes to counteract a particular evasive manoeuvre by the female. As she gets closer to egg-laying, the distance at which she begins to evade contact from the male decreases – so that, early in the mating season, the male following in a straight line after the female is most common, then later, following the female's zigzags are the most common, and then, finally, circling becomes the most common path. The decreasing distance at which the female begins to evade the male likely reflects her increasing sexual motivation, and differences across pairs may reflect differences in how long the animals have been pair bonded. It should be noted that copulation is preceded by the male successfully contacting and pecking the female's rump feathers; she then stops, after which the male shifts his beak contact to the top of her head, and, from this configuration, she squats and he mounts her (Pellis, 1982). Thus, abstracting the circling behaviour as a feature of male courtship in this species is misleading and is not an accurate reflection of the underlying organisation of behaviour.

In Cape Barren geese, circling by the male arises as he attempts to gain bill-to-rump contact with the female and she rotates to block that contact. Consequently, as the female rotates, the male's movements counteract her movements, leading to the pair maintaining a constant relationship between their bodies (Figure 2.4a). The circling in the courtship of another species of water bird that we have observed, a species of African ibis called the waldrapp, contrasts markedly with that of the Cape Barren goose (Pellis, 1989). The male waldrapp approaches the female, who squats when he is about a body length away. He then rotates to oppose her with his flank and proceeds to walk around her. Therefore, for the male waldrapp, the circling is determined by his behaviour not that of the female's. Moreover, as the male circles the female, the relationship between his body and that of the female is continually changing (Figure 2.4b). For

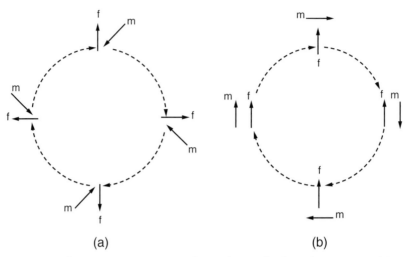

(a)   (b)

Figure 2.4 Schematic representation of courtship in the Cape Barren goose (a) and the waldrapp (b) shows that, while both males travel a circular path in space, their relationship to their respective female is very different. Adapted with permission from Pellis (1989)

the goose, the relevant CV is the bill-to-rump orientation, whereas for the waldrapp, the relevant CV is the relative distance to the female's body. In both species, the circle traced by the male may be what is most readily observed and easily measured by humans, but it is a poor behavioural marker as it does not reflect the different mechanisms that give rise to the circling of the two species.

Body Targets as a Starting Point for Identifying CVs in Inter-Animal Interactions

As the case of the geese shows, circling is a by-product of the male making compensatory movements to overcome the female's defensive actions as he manoeuvres to contact her rump with his bill (Figure 2.3). The key to understanding the organisation of the male's movements in space (e.g., straight line, zigzagging, circular) is his attraction to the female's rump. That is, the rump is the 'target' on the female's body that acts as the CV for the interaction. As in this case, many inter-animal interactions can be fruitfully studied by first determining the targets that are pursued by the contestants. The history of studying conspecific combat offers insight into the value of identifying such targets.

For obvious reasons, after the Second World War, there was much research interest in understanding the biological bases of aggression (e.g., Archer, 1988; Brain, Parmigiani, Blanchard & Mainardi, 1990; Huntingford & Turner, 1987). Whether combat was observed in free-living animals or in staged encounters in the laboratory, the problem was to identify appropriate 'behaviour patterns' that could then be quantified and analysed to determine how combat is organised and what controls the escalation to injurious contact. Based on the assumption that, in non-human animals, escalation to actual injurious contact is rare, such escalation was viewed as resulting from a failure to communicate effectively (Archer & Huntingford, 1994). However, as seen in the example of mating in Cape Barren geese, although circling by the male was originally interpreted as a courtship display, close analysis of the correlated movements by the partners revealed that the circling is a by-product of the compensatory manoeuvres of the two animals. This is unlike the case for the waldrapp in which the circling is a feature of the male's movement independent of that of the female (see Figure 2.4) and so is more likely to be a signal. During combat, because the two animals involved may be continuously countering one another's moves, there is an even greater risk that human observers may mistakenly interpret combat manoeuvres as signals (Blanchard & Blanchard, 1994; Geist, 1978; Pellis, 1997). Analysing interactions from the perspective of identifying targets that may be the CVs can help in resolving this descriptive problem (Pellis & Bell, 2011; Pellis & Pellis, 2015).

For the fighting of rats, we will focus on two supposed signals that have been identified: the adoption of a supine position that has been interpreted as a submissive posture and that of standing in a lateral orientation with an arched back that has been interpreted as a threat posture (Adams, 1980; Barnett & Marples, 1981). By lying on its back, the performer is signalling submission and, by remaining immobile, it is avoiding triggering a response from its opponent (Figure 2.5a). Similarly, by approaching an opponent in a lateral orientation, with an arched back and its hair standing on end (piloerection), an attacker appears to be threatening its opponent with imminent attack (Figure 2.5b). When these postures are coupled with knowledge of the targets attacked during fighting and by the correlated movements between the opponents, very different interpretations emerge.

(a)

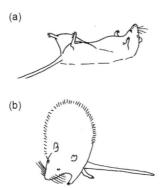

(b)

Figure 2.5 Drawings show two behaviour patterns that are associated with aggression in rats, supine defence (a) and lateral threat (b).

A common way to stage fights between adult male rats is to place an unfamiliar rat (intruder) into the home cage of another (resident). The resident initiates agonistic attacks by directing bites to the intruder's rump and lower dorsum, and the intruder manoeuvres to block these bites and may retaliate with ones directed at the resident's face (Blanchard & Blanchard, 1990). When the resident approaches from the side or to the rear of the intruder, and then lunges to bite its rump, the intruder can rotate, cephalocaudally, to lie on its back and so withdraw the targeted area of its body, inhibiting further attack (Figure 2.5a). In the same paradigm, house mice rarely use the supine posture (Brain, 1981), which can be explained by the resident switching from directing bites to the rump and instead, directing bites to the intruder's exposed underbelly (ventrum) (Blanchard, O'Connell & Blanchard, 1979). That is, the mouse's adoption of a supine submission signal could be viewed as being an ineffectual act of communication.

However, inspection of the fights involving the supine posture in rats shows that the intruder not only withdraws its rump from the oncoming teeth of the resident but may also lunge at the resident, directing a bite to its face (Blanchard, Blanchard, Takahashi & Kelley, 1977). Similarly, inspection of fights by mice reveals exactly the same pattern: the defender protects its rump by rotating to supine and retaliates by lunging toward its attacker's face (Pellis, Pellis, Manning & Dewsbury, 1992). What is different between mice and rats is that once the mouse defender rotates onto its back and then lunges, it quickly rights itself and runs away. That is, the mouse does not remain motionless in the supine position. The use of the supine manoeuvre is just as frequent in mice as it is in rats, but the use of the supine posture as a signal of submission is not. Given that the defensive supine tactic in mice is brief, lasting only a fraction of a second, whereas the supine submission signals in rats may last many seconds or even minutes, the latter is more readily noticeable to a human observer, especially if relying on the naked eye to score the posture. Frame-by-frame analysis of filmed sequences is needed to score the defensive supine tactic reliably, especially in the fast-moving mice (Pellis, Pellis, Manning & Dewsbury, 1992). The role of targets in organising interactions and the emergence of characteristic postures are even more graphically illustrated by the lateral posture.

In serious fights in male rats, a common defensive tactic adopted by the intruder is to stand on its hind feet and face the attacking resident, which often adopts the lateral posture (Figure 2.5b) as it approaches (Blanchard, Blanchard, Takahashi & Kelley, 1977). The advantage to the intruder in adopting this upright posture is that it keeps its rump and lower dorsum out of immediate reach, and, at the same time, places its teeth in an orientation to lunge at the resident's head if it reaches around for a bite. This dual defensive tactic by the intruder poses a problem to the resident as it attacks – reaching the intruder's dorsum while avoiding receiving a retaliatory bite to its own face. The lateral posture offers a solution to this dilemma. As shown in Figure 2.6, the intruder is standing on its hind legs at the rear, facing the approaching resident, which gradually adopts the lateral posture as it closes the distance. The resident then presses against the intruder's ventrum with its flank, but by being in the lateral orientation, maintains its head out of reach of the intruder's teeth. If, by pressing forward, the resident manages to push the intruder off-balance, the attacking resident then lunges to bite its exposed dorsum. The resident therefore keeps its head protected while creating an opportunity to launch a biting attack at the intruder's target. This lateral manoeuvre is common to many rodents, with comparative analyses showing that the lateral orientation serves this dual protective and offensive function. Another feature of the lateral manoeuvre is that the body is arched – a posture that is achieved by the animal bringing its hind feet and forefeet closer together. Having its four feet closer together enables the attacking animal to swerve towards or away from its opponent more quickly than if its feet were to remain further apart. Only its third feature, the piloerection of the fur, is

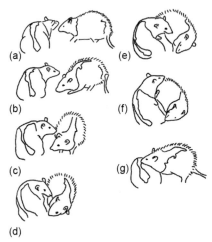

Figure 2.6 The lateral combat manoeuvre in the fighting of adult male rats is illustrated. By sidling up to the upright defender in a lateral orientation (a), the attacker can press into the intruder's ventrum with its flank (b) while keeping its own face out of reach (c–e). If this pushing succeeds in off-balancing the intruder (f), the attacker will then swing around and bite the defending intruder's lower flank (g). Reprinted from Pellis and Pellis (1987) with permission (Copyright © 1987 John Wiley & Sons, Inc.)

not an essential part of the lateral manoeuvre as a combat tactic (Pellis, 1997).

That the lateral posture is best interpreted as a combat tactic rather than as a threat signal is supported by what happens when the defending animal manages to deliver a successful retaliatory bite to its opponent's face. For example, in an encounter between a resident–intruder pair of grasshopper mice, after several attacks by the resident involving the use of the lateral manoeuvre to gain access to the intruder's dorsum, the intruder managed to deliver a bite to the side of the resident's face. While before this bite was delivered, the approaches had the classic lateral orientation, the ones immediately after changed dramatically. The resident began by approaching laterally, but when it was within a body length away it pivoted, moving its head further away from the intruder, so that when body contact was made the resident was oriented with its rump facing the intruder. In one case, this 'rump push' succeeded in pushing the intruder off balance, but because of the distance of the resident's head from the intruder's body, by the time the resident swerved around to lunge at the intruder's exposed dorsum, the intruder had managed to regain its footing and run away (Pellis & Pellis, 1992).

The above example shows that the lateral orientation is a compromise between the resident attempting to deliver a bite to the intruder's dorsum while simultaneously avoiding being bitten on its own face. Moving one's face too close or too far away from the defending animal undermines the success of this tactic. Therefore, placing the postures most apparent to human observers (Figure 2.5) in the context of the targets that are attacked and defended and the moves and countermoves made by the opponents, two actions which seem to be signals are shown to be combat tactics; although in some cases, one or both may also serve a communicatory function (Blanchard & Blanchard, 1994; Pellis, 1997; Pellis & Pellis, 2015). In these encounters, the targets constitute the perceptions that guide the actions of the participants, and so much of the behaviour by both animals is best interpreted as being compensatory to the actions of the opponent that disrupt those perceptions (Bell, 2014; Pellis & Bell, 2011). The lesson is that the 'target as CV' provides a useful starting point in analysing interactions that provide insights into the organisation of the behaviour (Pellis & Bell, 2020).

Solitary Behaviour Poses Novel Challenges

As pointed out to us by Ilan Golani (personal communication, 1985), despite the dynamic flurry in interactions between animals, it can be assumed from the start that the behaviour of one animal is going to be important in shaping the actions of the other. It is highly likely that some facet(s) of the inter-animal relationship is to serve as the perception that is being controlled (e.g., Golani, 1976, 1981). For solitary behaviour, however, particularly an animal moving around an empty, open space, no such self-evident cues are available to the human observer. However, detailed studies of how rats and other rodents navigate an open space have shown that, here too, animals create meaningful relationships with their surroundings. From a descriptive point of view, making sense of those relationships can provide a deeper understanding of the overt behaviour than by arbitrarily scoring aspects of movement in space. This has been beautifully revealed by studies showing how rats come to know a new area and how their behaviour changes in that space over time.

When placed in an unfamiliar open field, a rat will remain motionless; then, after some time has elapsed, it will start moving. These movements unfold in a particular pattern. First, the animal moves its head from side to side, and, with proceeding movements recruits more of its body, making increasingly larger movements until it makes a complete circle pivoting around its pelvis. Second, the same procedure occurs in the forward dimension, until the rat steps forward. Third, the rat makes a similar progression vertically until it stands upright on its hind feet. Moreover, as the expansion of movement in these three dimensions overlap, inter-mediate combinations may arise, such as walking in a circle (combining lateral and forward movements) (Eilam & Golani, 1988; Golani, Wolgin & Teitelbaum, 1979; Szechtman, Ornstein, Teitelbaum, & Golani, 1985). In part, this unfolding pattern serves to translate the animal's own body movements into a map of the surrounding space. The movements in the new space gradually become more expansive, until the animal moves through all or nearly all the available space.

However, not all space is treated in the same way by a rat. First, in an open field enclosed by a wall, the manner in which the open area of the enclosure is explored is different to when a rat confronts an open space that is not surrounded by a wall. The reason for this is that, rats, like other small rodents, are thigmotactic (Barnett, 1975) – that is, they will tend to move so as to retain tactile contact with a vertical surface. Consequently, in the presence of a wall, rats and mice will gradually make forays further and further into the central area, but will always return to the safety of the wall (Golani, 2012).

In the absence of a surrounding wall, there is no safe haven. However, as the rat moves about, it gradually creates safe havens in the open area around which its further movements in the larger area are organised. After several steps that may traverse a number of body lengths, the rat begins to stop at regular places. Some of these stopping sites become 'home bases' from which a rat will make forays (Eilam & Golani, 1989). The animal may start at one home base, then traverse open terrain and stop at another home base. Interestingly, after home bases are established, it is only at these locations that the animal will engage in behaviours such as resting, rearing, and grooming. That is, the movements in the open field, including the direction of movement, the speed of movement, and the performance of outwardly and self-directed behaviour, all begin to make sense within the context of the value attributed by the rat to specific locations in this seemingly barren and uniform two-dimensional arena.

Understanding the sensory information – be it visual, olfactory, vestibular or proprioceptive – that is used by the animal to identify the spatially relevant perceptions further helps us make sense of the particular behavioural actions performed (Wallace, Choudry & Martin, 2006). That is, the rat has a dynamic interaction with its physical environment, and this creates a structure within that environment. In this context, simply scoring the overt acts of the animal – such as the frequency of rearing and the amount of space traversed – would miss much (Eilam & Golani, 1990). As we have seen, in understanding inter-animal fighting or in open-field behaviour, a description that is sensitive to the dynamic interaction of the animal with its surroundings can provide insight into the organisation of the behaviour studied. Moreover, a descriptive framework that incorporates that dynamic exchange can often explain more anomalies than one that, from the outset, just focusses on scoring readily observable patterns.

Testing Hypotheses

Once a putative CV is identified, it needs to be tested as to whether the animal truly controls that perception. For example, the dynamics of interaction between two male rats fighting strongly suggest that the rump and lower dorsum are the targets of attack and defence (Pellis & Pellis, 1987). Additional evidence supports this conclusion. First, when an intruding rat is anaesthetised and placed into a resident's home cage, it is attacked by the resident and bitten. If the intruder is placed on its back, the resident burrows beneath the intruder to bite at the intruder's dorsum (Blanchard, Blanchard, Takahashi & Kelley, 1977). That is, the resident bypasses other body areas available to it and manoeuvres to gain access to the hidden lower dorsum and rump. Second, among rodents such as mice and rats, odour is an important cue as to the presence and identity of conspecifics (Brown, 1985), and indeed, peripheral anosmia (blocking olfaction) prevents attacks on unfamiliar intruder males (Alberts & Galef, 1973). In montane voles, it is the odour that emanates from the opponent's hip glands that attracts bites.

Removal of the gland on one side of the vole leads to more attacks on the side with the remaining gland (Jannett, 1981). The montane vole illustrates two important lessons about CVs.

First, it is rarely the case that the CV we can detect is actually the perception that is controlled by the animal. The compensatory actions by the male goose indicate that he is keeping the relationship between his bill and female's rump constant. It cannot be ascertained, however, from this constancy, as to the exact perception that is relevant to the male goose – after all, the bill–rump association may be correlated with the actual perception that is controlled (Bell & Pellis, 2011). Experimental evidence, such as removal of one of the scent glands in a male montane vole, can provide clues as to what the actual perception may be that is critical to the animal. Second, to understand why particular CVs are defended in a particular species, other aspects of the animal's biology and evolutionary history need to be taken into account (Pellis, 1997). These issues are beyond the scope of the present book, but, from a methodological perspective, knowing the potential CVs involved can make sense of much of the overt behaviour noticeable to a human observer.

The 'target as CV' posits that the behaviour performed by the participants can be accounted for as being compensatory actions to the disturbances created by the opponent in gaining access to or protecting that target. This approach has been fruitfully applied to explaining many of the actions performed during conspecific aggression, amicable interactions between conspecifics and during predation (e.g., Blanchard & Blanchard, 1994; Geist, 1978; Pellis, 1997; Pellis & Officer, 1987). However, what also needs to be tested is whether all aspects of the behaviour performed can be explained as being compensatory.

Tinbergen and Lorenz (1938) described how a goose retrieves an egg that has rolled out of its nest. When an egg is outside the nest, the goose – sitting in the nest – stretches out its neck, places the ventrum of its bill over and in front of the egg and then pulls backward, thus moving the egg towards the nest (Figure 2.7). If contact between the beak and the egg is broken, by either the egg accidently rolling to one side, or because an experimenter has slyly pulled it to one side, the goose continues the backward movement of its head until it reaches the anterior edge of its body and only then will the goose raise its head, scan the surroundings, locate the the egg and then move its head forward to contact the egg again.

(a)   (b)

(c)   (d)

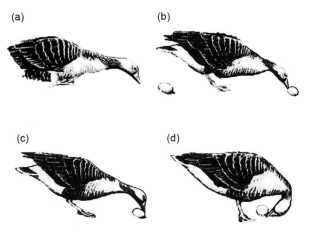

Figure 2.7 A sequence of drawings shows a greylag goose retrieving an egg. It reaches out with its neck (a) until it makes contact with its beak. (b) The goose then shifts its contact so that, with the anterior ventral surface of its beak, it contacts the far side of the egg. (c) From this configuration, by contracting its neck, the goose pulls the egg back towards its nest (d). Adapted from Tinbergen and Lorenz (1938) with permission (Copyright © 1939 Blackwell Verlag GmbH)

This completion of the backward action before a new retrieval attempt is made is what drew Tinbergen and Lorenz to the conclusion that the action itself is a fixed motor unit (a fixed action pattern), which has to be completed before another is re-started. Looking at the clips of egg retrieval available online from YouTube from the perspective of PCT, a very different conclusion about why the goose 'completes' the backward movement was drawn (Marken, 2002).

According to Marken, the goose continues to move its head backwards to 'search' for the missing egg and it is only when it reaches its own body and does not find the egg that the goose looks around, locates the egg and starts again. That is, the bill–egg contact is the CV (the perception that is controlled) and, on losing it, the bird makes backward compensatory movements to re-establish the perception. Thus, we have two contrasting hypotheses to account for the goose's head movement backwards in the absence of the egg. In the Tinbergen and Lorenz hypothesis, the backward movement is a fixed motor action ('the motor action hypothesis'), whereas in the Marken hypothesis, the backward movement is a compensatory action used by the goose to regain the lost contact ('the search hypothesis'). To evaluate these two competing hypotheses, we need to explore the behaviour of the goose in detail, both when the egg is present and when it is missing.

In the original description, two components of the goose's behaviour were recognised – the backward retrieval movement and small, side-to-side head movements. The former is an invariant action that pulls the egg toward the goose's body and the latter is a variable action that maintains a straight-line path of the egg to the body (McFarland, 1971). The oblong shape of the egg results in the egg veering to one side or the other, so the goose uses the sensory feedback on its bill to move its head from side-to-side to maintain the egg's linear path. Indeed, the side-to-side movements cease once contact with the egg is lost. This dichotomy between an invariant retrieval movement (pulling the head backwards) and variable side-to-side head movements is misleading. The distance of the egg from the nest can vary from one case to another and nests may be at a slightly different height from the surrounding ground. Consequently, the magnitude of the goose's forward neck extension and the postural adjustments it makes with its body and legs are likely to differ from one case to another. Depending on the context, different combinations of muscle contractions would be needed and so, even the supposedly invariant component – the goose's extension and retraction of its neck – contains variability. (For a detailed exploration of this issue in another behaviour, see Pellis, Gray & Cade, 2009.) Therefore, the whole action, forward and back, may be better thought of as a functional cycle – to locate and retrieve the egg – with all parts of the action being variable, as compensatory action taken to offset disturbances to the bill–egg relationship (Pellis, Pellis & Iwaniuk, 2014; von Uexküll, 1934/2010). In this way, the search hypothesis seems stronger, making sense of all phases of the egg retrieval, including the goose continuing the backward movement of its neck as it searches for the missing egg. But there are problems that are not resolved by this hypothesis either.

First, let's examine the search hypothesis more closely. The movement backwards occurs so as to reconnect the goose with the egg, but why is this movement directly backwards? The reason for the side-to-side movements of the goose's head is to keep the egg from rolling to one side or the other: without this 'herding' with its bill, the egg could roll in any direction. We tested this by rolling uncooked chicken eggs on our kitchen floor. With a

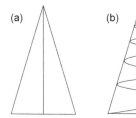

Figure 2.8 Because of its oblong shape, a rolling egg is unlikely to continue to move in a straight line. Rather, the further the egg moves, the more likely it is to veer off that path. The isosceles triangle shows the potential space occupied by an egg as it rolls. As the goose moves its beak directly backwards towards its body (the straight line from top to bottom), the chance that it contacts the egg diminishes (a). A more effective movement strategy would be for the goose to sweep, with its beak, from side to side, with the sweeps growing in magnitude as it moves the beak towards its body (b).

slow-to-moderate push, one that mimics the goose reasonably closely, the average distance travelled before the egg deviated from its linear path was 11.2 cm (range: 5–35). The average angular deviation was 59.4° (range: 20–90) and the average distance travelled in that new direction before stopping was 19.6 cm (range: 6–50). Giving the egg a harder push increased the distance travelled before its path deviated to an average of 28.0 cm (range: 10–63). Only pushes that far exceeded what would happen in the real situation produced the occasional roll that remained in a straight-line path until the egg stopped (or hit a wall and broke). At more reasonable speeds of travel, the further the egg moved, the more likely it would be for it to veer off from the straight-line path. Depicting the possible space that the egg could occupy as an isosceles triangle, with the apex being the egg's starting point and the base being the location of the goose's body, reveals that a goose moving its bill backwards in a straight line would be highly unlikely to make contact with the egg (Figure 2.8a). Indeed, in the real world of the area around a nest, irregularities in the substrate, such as a small clump of grass, could unexpectedly stop an egg in its tracks. A better search strategy would be for the goose to make ever wider side-to-side sweeps of its head as it moves its head backwards, thus increasing the likelihood of contact with the errant egg (Figure 2.8b). Such non-linear paths have been reported for birds in other situations, suggesting that a more efficient search pattern is possible. For example, wandering albatross use olfaction to locate their prey, fish that are swimming close to the surface of the ocean. However, the odour from their prey disperses laterally and downwind; consequently, the albatross fly crosswind to detect a scent and then zigzag to locate the source. Modelling alternative manoeuvres suggest that zigzagging is the most efficient foraging strategy (Nevitt, Losekoot & Weimerskirch, 2008).

Indeed, since the goose raises its head to locate the missing egg visually, once the backward movement results in its head reaching its chest, why does the goose not simply scan visually for the displaced egg once contact is lost rather than continue the backward movement that is unlikely to ensure that it regains contact with the egg? There is another problem with the search hypothesis in explaining the goose's egg retrieval behaviour. Close examination of actual sequences of the backward movement of the goose's head in the absence of an egg shows that the movement does not stop when its bill reaches the base of its neck and chest, but, rather, the goose makes an upward tucking movement on the lateral edge of either side of its body. It is not clear how this could be a searching movement. It occurs when the egg is present and seems to be used by the goose to press the egg firmly against its body, before it tucks the egg underneath its body. And yet, here it is, in trial after trial, produced in an invariant manner, a motor action seemingly performed for its own sake and not as part of a compensatory movement to re-establish egg contact.

So, while there are deficiencies in the hypothesis that the retrieval action is an invariant action pattern (Tinbergen & Lorenz, 1938), the search hypothesis is also deficient. That is, not all behavioural actions can be explained as compensatory manoeuvres for disturbances to salient perceptions (CVs). Chapter 3 examines factors that regulate the intrinsic organisation of motor actions, and how such organisation constitutes another principle that needs to be considered when deciding on selecting behavioural markers for quantitative analyses.

3   The Deep Structure of Behaviour

> That behavior is organized in functional motivational systems lies at the core of the set of ideas constituting the behavior systems perspective.
>
> (Burghardt & Bowers, 2017, p. 342)

Bats hunt moths using sound as a means of echolocation to track down their prey (Acharyra & Fenton, 1992). The same high-pitched sounds that can indicate the presence of a bat can elicit a very different defensive response by various species of moths. Those with camouflage colouration alight on a suitable tree trunk and cling closely to the substrate, whereas those with brighter colouration that makes them easier to spot when close, fly away (von Uexküll, 1934/2010). This example shows that knowing what an animal can sense and which of those sensations are perceptions that guide behaviour may not be sufficient to explain the behaviour associated with those perceptions. For example, why are the feathers at the base of the female's tail an attraction to the male Cape Barren goose? As noted in Chapter 2, answering such a question may require broad, comparative research that is able to trace the history of the trait, but sometimes, we may just have to shrug our shoulders and accept that the answer may never be found. Nonetheless, even if we do not know why the female goose's tail feathers are an attraction, knowing that they are enables us to account for much of the courtship behaviour of the male goose. In other cases, more information on the organisation of the behaviour can provide clues as to why some actions are produced and others are not.

The flick of the head to the side and up by the goose as it completes the backward journey of its head in egg-rolling illustrates that many movements are not strung together into functional actions in an arbitrary manner but rather, are causally linked in a particular sequence. Quite simply, many movements do not come as atomised units that can, willy-nilly, be combined together (Flash & Hochner, 2005; Kolb & Whishaw,

2015; Skinner, 1938), but instead, are combined into actions in highly distinguishable ways, the expression and coordination of which may be controlled by highly specific environmental stimuli and neural circuits (Baerends, 1976; Tinbergen, 1951). Again, righting provides an illustrative example.

As we have already seen, when inverted and dropped in the air, a cat rotates cephalocaudally so that when it lands, it is prone (Magnus, 1926). If the cat is blindfolded, the error in the position of its head is detected by its vestibular system. But once the cat rights its head to prone, what leads to the rest of its body rotating to prone? One explanation posited was based on the concept of the chain reflex (Sherrington, 1906). That is, the vestibular signal causes the cat's head to rotate, which, in turn, causes its neck vertebrae to move against one another, producing a proprioceptive signal, which then triggers its upper body to rotate. In turn, the rotation of its upper body causes the vertebrae of its lumbar area to move against one another, producing another proprioceptive signal that then triggers the rotation of the cat's lower body. Thus, the cephalocaudal rotation of the whole body involves a sequence of chain reflexes (Magnus, 1924). However, using a high-speed camera (S. Pellis, Pellis & Teitelbaum, 1991), we filmed rats righting in the air and saw some movement patterns that were inexplicable from the perspective of the chain reflex.

As the rat's head and shoulders began to rotate in one direction, say, to the right, its lower body and tail began to rotate to the left. These observations contradict the chain reflex explanation in two ways. First, the rat's upper and lower body began moving simultaneously, and second, the direction of that movement is conflicting, not complementary. The counter-rotation of the rat's lower body provides a mechanical anchor by which its forequarters rotate to prone. Once the rat's forequarters are fully prone, they then provide a mechanical anchor for its hindquarters to reverse the direction of righting – in this case, from left to right, so that they also achieve the prone position. These counter-rotations by different body parts are among the strategies used by which to provide torque when there is no substrate against which to push (Jusufi, Zeng, Full & Dudley, 2011). Of more direct relevance to our current considerations, this simultaneous and opposing movement of the upper and lower body shows that righting involves a coherent and coordinated concatenation of body movements, and is not a fragmented sequence of movements (Pellis, 1996). The importance of

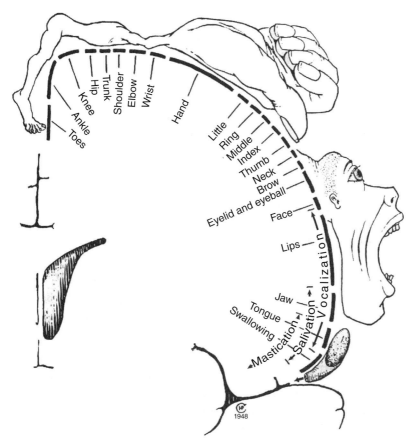

Figure 3.1 A representation of the motor homunculus shows that different portions of the motor cortex are associated with different areas of the body. Those areas of the body with more muscle control (e.g., hand vs. foot) have a greater share of the cortex (i.e., more neurons). Adapted from Penfield and Rasmussen (1950) by Graziano (2016) reprinted with permission (Copyright © 2016 Elsevier Ltd.)

packaging together movements by different parts of the body is well illustrated by the neural control of actions.

The motor homunculus (Figure 3.1) is a classical depiction of how movements are supposed to be controlled in an atomistic manner by the mammalian motor cortex (Penfield & Boldrey, 1937; Penfield & Rasmussen, 1950). The construction of this model was derived using electrical stimulation of a short duration (50 ms or less) to different regions of the motor cortex, revealing muscular contraction in different areas of the body, with more neural tissue contributing to the body parts that have greater motor flexibility and control (e.g., the human thumb vs. the human big toe). This approach also revealed that, within an area of cortex linked to a particular body part, such as the thumb, electrical stimulation of some cells leads to abduction while stimulation of others leads to adduction, some to flexion and some to extension. Movements are atomised at the level of particular muscular contractions and then combined into specific actions.

Using this procedure with rhesus monkeys, Michael Graziano accidently stimulated the animal for longer (0.5 s), and, to his surprise, found that this produced discernible whole-body actions that coherently packaged together movements from multiple, connected body parts (e.g., fingers, hand, lower arm, upper arm) (Graziano, 2009, 2016). Following up on this chance discovery with systematic testing of different areas of the motor cortex revealed that stimulation of different cells led to different, functional actions that are the same as those used in normal daily life (Figure 3.2). Subsequent research has extended these findings not only to other primates (Kaas, Gharbawie & Stepniewska, 2013) but also to rats (Brown & Teskey, 2014).

Older literature that involved electrical stimulation of neural centres below the cortex, in the midbrain, especially around the hypothalamus, not only showed that coherent multi-segment actions could be activated but also that whole sequences of actions are linked together in a functional manner. Electrical stimulation of one part of the hypothalamus of chickens can elicit a whole sequence of actions associated with anti-predator behaviour, starting with the chicken becoming alert, to squawking and pacing, to defecating and then, with increasing electrical stimulation to the maximum, flying away (von Holst, 1973). Stimulation of other areas of the hypothalamus or lower brainstem can elicit sequences of predatory behaviour in carnivorous mammals (MacDonnell & Flynn, 1966). That is, movements are not independent of one another, but are linked together in articulated packages that may form a behaviour system that has a hierarchical arrangement in its constituent movements, actions, and sequences (Baerends, 1976; Tinbergen, 1951). This 'behaviour system' conception has been developed into various theoretical frameworks (e.g., Hogan, 2001; Panksepp, 1998; Teitelbaum, 1982; Timberlake, 2001), but what they all have in common is the view that few behaviour patterns are

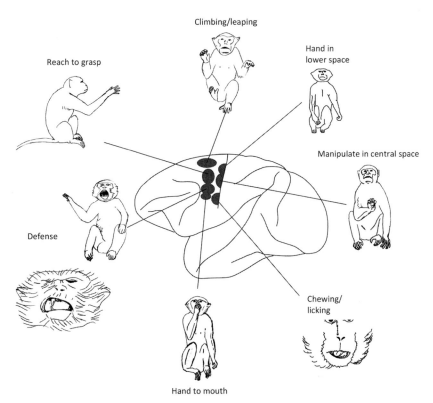

Climbing/leaping

Reach to grasp

Hand in
lower space

Manipulate in central space

Defense

Chewing/
licking

Hand to mouth

Figure 3.2 The differing action patterns elicited by prolonged stimulation
of different areas of the motor cortex of macaque monkeys. Reprinted from
Graziano (2016) with permission (Copyright © 2016 Elsevier Ltd.)

independent, stand-alone actions (Bowers, 2017; Burghardt & Bowers, 2017). This has implications for what we choose to measure.

Recall from Chapter 1 that the time-to-right measure turned out to be an inadequate way of scoring the development of righting. Given this inadequacy, we could choose a constituent action that is central to righting – the rotation of the neck and/or upper body around the longitudinal axis of the body – to determine when righting begins to mature. But again, this would prove inadequate. At an age when trigeminal stimulated righting recruits neck and shoulder rotation, upper body tactile stimulation fails to do so, but instead, recruits forelimb movements to flip the body over (V. Pellis, Pellis & Teitelbaum, 1991; Pellis, Pellis & Nelson, 1992).

Even though in the mature, adult form, both types of tactile stimulation recruit upper body axial rotation, in early development, access to rotating the axial musculature is not available to both types of righting. Knowing that the ability to perform the movement exists is not sufficient to account for the behaviour. Therefore, unless the precise context of usage is well known and described adequately, using axial rotation as a proxy behavioural marker for righting can be as inadequate as using the time-to-right measure. The reason is that axial rotation of the upper body is associated differently with the organisation of trigeminal and upper body tactile-triggered righting.

A Bite is Not a Bite

You have likely seen, either in real life or in a documentary, a cat catch a mouse. The cat detects a mouse in the distance then proceeds to stalk it, with its body low to the ground and its every step slow and deliberate. Once the distance between the predator and the prey is closed, the cat pounces, pins the mouse's hindquarters with its forepaw and delivers a bite to the nape of the mouse's neck. Biting the nape of the mouse's neck is central to the cat's predatory behaviour, but the biting of the nape is not limited to its prey. During their breeding season, cats can make a racket in your back lane or garden, as males compete for access to a receptive female. Once the males establish, by fighting, which one of them has priority access to the female, the couple will then court, and, if she deems him worthy, the female cat will position herself in front of the male, squat and then raise her hindquarters (i.e., adopt a lordosis posture – her back arches so that her rump is elevated and her tail is deflected to one side, exposing her vagina), at which time the male will mount and copulate with her. During copulation, the male bites the nape of the female's neck and maintains this grip until he ejaculates. In cats both small and large, biting the back of the neck is a reliably documented behaviour in both predation and sex (Leyhausen, 1979). What distinguishes between the two bites – the bite to a prey compared to one to a sex partner – is that, when delivered to a mouse or a rat, the cat applies enough pressure for its canines to pierce the skin and damage the prey's spinal cord, whereas when the bite is applied to a female cat, the grip is firm but excessive pressure is inhibited so that lethal damage is not inflicted.

One way in which to conceive of these two bites is that the cat's bite is the same in each case, but with a difference in the degree of pressure applied. In this view, biting is an independent action pattern that can be recruited and applied with suitable modification to different behavioural contexts – but this is not the case. Damage to one brain area renders a male cat able to deliver lethal bites to the neck of a mouse but unable to bite the nape of the neck of a female cat, while damage to a different area of the brain leads to a male cat that can bite the neck of a female cat but not the neck of a mouse (Hunsperger, 1983). There is differential access to the two types of biting – predatory or sexual. This distinction is not just of theoretical interest but is one on which Mabel Stark staked her life.

Mabel Stark was a tamer of large cats for the Ringling Brothers and Barnum and Bailey circus in the 1920s (Hough, 2001). After having lions and leopards perform a variety of feats, her final act was with a large male tiger named Rajah. After directing Rajah to jump through hoops, roll over and do other tricks, Mabel ended the act by turning to the crowd and bowing, which meant that her back was facing Rajah. At this moment, Rajah pounced on her back and delivered a bite to the back of her neck. The audience gasped. Undaunted, Mabel struggled, broke free and sub-dued Rajah. The crowd went wild – Mabel had survived a predatory attack of an adult tiger! No wonder she was such a prized performer! However, unnoticed by her audience, between finishing the first part of her act with the lions and leopards and the final part with Rajah, Mabel had a short break, and had returned to the arena wearing a white jumpsuit. Understandably, as performing her act with the big cats was a dusty and sweaty one, Mabel coming back in fresh clothes was not seen by the crowd as untoward. What the crowd did not notice was that her white jumpsuit hid something that Rajah left behind in his attack on Mabel – his ejaculate. Mabel did not survive a predatory attack, but a sexual one. It was not a killing bite, but a love bite, and a feline love bite does not penetrate the skin and produce a bloody wound as does a killing bite. Different bites are associated with different behaviour systems. If it were otherwise, Mabel's career may have been short-lived.

She was luckier than she knew, as the act counted on a 'top-down' activation of biting. That is, the tiger's sexual motivation activates the behaviour patterns and sequence that leads to copulation and the associ-ated love bite. However, sometimes, the connections can go in the opposite direction. For example, Lind (1959) reported that, if, when diving, a male duck configures his body into the posture that occurs at the onset of courtship, he may begin a courtship sequence rather than complete the dive. Closer to home, try smiling when you feel miserable and you will feel happier. In one experiment, human subjects were asked to hold a pencil in their mouth in one of two ways. They were asked to hold the pencil either with their front teeth, resulting in a grin, or with their lips, resulting in a pout. The former group of subjects reported being more amused than the latter group when shown humorous cartoons (Strack, Martin, & Stepper, 1988). While there is controversy about the strength of these gesture-induced effects on emotions (Petri & Govern, 2004), the empirical evidence does suggest that, under some circumstances, putting on a particular expression may influence the way one feels (Han, Luo & Han, 2016; Niedenthal, 2007; Rychlowska et al., 2014). In these cases, there is a 'bottom-up' relationship between performing a specific behaviour and the brain mechanisms involved. That is, by performing the action, the central, motivating mechanisms associated with that behaviour are activated. What if the bite delivered by Rajah negated his sexual desires and activated his predatory instincts? The crowd at the circus would have had even more reason to gasp, as the white suit would have made the blood oozing out of Mabel even more evident.

Beware of End Point Measures

Whether top-down or bottom-up, what the previous examples illustrate is that behavioural actions are linked together through deep organisational structure and so simply scoring the overt actions without understanding that structure can be misleading. For example, when reaching out to grab something to eat, the point of the action is to place some food in your mouth. A simple way to score such an action numerically is to count the number of times food is reached for and divide this number into the number of such cases that led to a successful delivery of food to the mouth, multiply by 100 and you have a numerical index of the rate of success. Laboratory rats, which are used extensively as stand-ins for humans in biomedical research, provide an example. Two things are often required for such testing: the measurements must be simple and highly reliable (recall

Figure 1.1) and a large number of subjects need to be processed. With these constraints in mind, consider evaluating the effects of brain damage, say, mimicking a motor cortex stroke in humans, on motor performance in rats. For such animal models, often their greatest utility is in assessing the benefits of rehabilitation therapies. In this case, does the original brain damage affect motor performance (i.e., success in reaching for and retrieving food is reduced) and does the therapy improve recovery (i.e., regain the level of success that was present before the damage)?

In one test that has been developed, rats are placed in an enclosure in which they face a wall of vertical bars through which they can reach with their forepaws to grasp chicken feed that is in a tray (Figure 3.3a). They are given unlimited opportunity to reach through the bars to retrieve food pellets, and by videotaping the trial the number of attempts to grasp food can be counted for each rat as can its success in retrieving the food pellets (Whishaw, O'Conner & Dunnett, 1986). This simple scoring system has the benefits of high inter-observer reliability and, by increasing the length of the enclosure, can test multiple rats simultaneously, and so increase the sample size. The effects of brain damage can be assessed for decrements in motor performance and if recovery occurs over time. If there were spontaneous recovery, the potential benefits of therapy on accelerating or increasing the level of spontaneous improvement can be assessed. In some experiments, the level of recovery is 100 percent. The implication of such a finding is that the animal regains the normal use of the behaviour that was lost due to the brain damage.

Rats perform a very particular sequence of actions to reach for, grasp and retrieve a food pellet (Whishaw & Pellis, 1990). The rat points its nose toward the object to be grasped, and then, shifting its body weight to one side of its body, allows its now freed forelimb to be lifted so that it can position its paw near the tip of its nose. The rat then abducts its elbow so that it is aligned near the midline of its body. As it does so, it rotates its lower forelimb so that the palmar surface of its paw opposes the side of its face. From this position, the rat extends its arm forward, thus moving its paw towards the food item. As it does this, the rat rotates its arm so that its palm is facing the ground. Similarly, there is a structured pattern for the grasp and the retrieval that brings the food item to the rat's mouth. The sequence was derived from detailed, kinematic analyses of the spatiotemporal movements of the interlinked limb and body segments (as illustrated

(a)

(b)

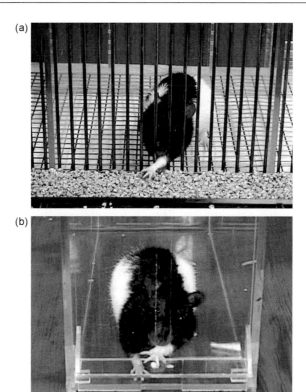

Figure 3.3 Testing for reaching success. (a) In the 'food-tray' task, the rat is able to reach through a slot to grasp food from a tray that is filled with food. There are multiple slots through which the rat can reach and if the food is dropped, it is lost through the floor of the cage. (b) In the 'single-pellet' task, a rat is required to reach for a single food pellet. There is a single slot through which the rat can reach to grasp the food. As the floor of the cage is Plexiglas®, any dropped food can be retrieved. Adapted from Vergara-Aragon et al. (2003) with permission (Copyright © 2003 Society for Neuroscience)

in Appendix A). To view this pattern, a single-pellet test procedure was used in which the animal was in a transparent, acrylic plastic (Plexiglas®) enclosure with walls and a single slot at the front facing a small, horizontal platform with a well containing the pellet (Figure 3.3b). The reaching can be videotaped from the side, the front, from above or below. These various views helped confirm the robustness of the various reaching components. Subsequent testing that included close-up video sequences of the paws grasping (Whishaw & Gorny, 1994a) and X-ray cinematography to reveal the movements of the skeletal elements (Alaverdashvili, Leblond, Rossignol & Whishaw, 2008), further validated the basic kinematic structure of the reaching in rats.

Armed with this understanding of how reaching in rats is structured, the recovery of reaching success in brain-damaged rats could be re-assessed. First, different types of brain damage disrupted different aspects of the reaching sequence (e.g., Whishaw, Pellis & Pellis, 1992; Whishaw, Pellis, Gorny & Pellis, 1991) and second, even rats that regained their pre-damage level of success often did not regain the pre-damage kinematic pattern (e.g., Whishaw, Pellis & Gorny, 1992a). That is, the rats may have recovered their ability to retrieve food items successfully, but they did not recover the original movement sequence. Rather than the improved reaching success reflecting recovery, the animals adopted compensatory strategies for the lost capacity (Alaverdashvili & Whishaw, 2013). Recovery and compensation are two very different processes, between which measuring reaching success cannot differentiate. As a caveat, it should be noted that, once the components of the reaching and retrieval sequence in the rat were identified and characterised, they could then be scored as 'typical', 'absent' or 'impaired', and then the scores for each component can be combined into a numerical index that can be used by many laboratories to assess motor impairment in rats (Whishaw, Whishaw & Gorny, 2008). Thus, while others can score the behaviour, rather than describe it, they can do so assured that the scores closely reflect the underlying organisation of reaching (Whishaw, 1996), thus achieving the coveted combination of accuracy and precision (see Figure 1.1c).

When Deep Structure Is Not Sufficient

One of the concerns about the structure of reaching in rats is that it may be context-dependent, whereby the constraints of the task force the animal to make particular movements. For example, the 'elbow-in' component (i.e., the abducting of the elbow towards the midline) of reaching may be due to the rat's forearm being forced to move through a narrow, vertical slot. To determine if this were so, rats were tested reaching toward food through a horizontal slot, in the absence of any barrier and when spontaneously eating. In these cases, the rats did not need to move their elbow in toward their midline but did so anyway (Whishaw, Dringenberg & Pellis, 1992; Whishaw, Pellis & Gorny, 1992b). Comparative studies showed that other rodents similarly moved their elbow in before extending their arms, whereas some non-rodent species do not (Ivanco, Pellis & Whishaw, 1996; Whishaw, Sarna & Pellis, 1998). Finally, rats with some forms of brain damage no longer brought their elbow in, but still managed to reach through a vertical slot (Whishaw, Pellis & Gorny, 1992a). Therefore, the elbow-in movement is not a by-product of the rat's body morphology or the context in which it is constrained to reach (Whishaw, 1996). Rather, it is a built-in bias as to how rats, and some other rodents, organise their preparatory arm movements before reaching forward.

The elbow-in movement in the reaching behaviour of rats underscores the main point of the principle explored in this chapter: some actions performed by animals arise from the non-arbitrary manner in which they are functionally organised. Again, as shown in the case of reaching, this bias can be eliminated by brain damage, clearly indicating that this bias is wired in the brain. Such motor biases are crucial as otherwise there are so many degrees of freedom of movement; think about all the joints in a vertebrate's body and all the potential movement around those joints – that putting together a coherent sequence of movement from scratch could be excessively time consuming and neurally expensive. Having built-in limits as to which actions and postures are adopted provides a cost-saving device to allow for actions to be performed more efficiently in specific circumstances (Lelard, Jamon, Gasc & Vidal, 2006; Muir, 2000; Pellis, 2011; Pellis, Pellis & Iwaniuk, 2014). However, the production of a con-textually invariant action cannot by itself be taken as evidence that the action must arise due to the peculiarities of neural wiring. That proposition, like the one discussed in Chapter 2, is a hypothesis and hypotheses need to be tested.

It is well known that horses change gaits – from walking to trotting, and from trotting to galloping – based on the speed of locomotion (Gambaryan, 1974). Experimental studies with cats, rats, and other vertebrates have shown that there is a pattern generator in the brainstem that is responsible for switching gaits, so that, with appropriate electrical stimulation, the subject can be induced to switch from one gait pattern to another (Grillner, 1975). Following brain damage involving depletion of the neurotransmitter dopamine, in the movement areas of the brain, rats are made akinetic (i.e., they do not spontaneously move). However, treatment with certain drugs can induce them to walk, but their gait is atypical, involving a diagonal pattern (right hind paw lands, left forepaw steps forward) rather than the typical lateral pattern (right hind paw lands, right forepaw steps forward). Some researchers interpreted this gait switch as one that involved changing the central pattern generator from one state to another. An alternative explanation emerged that did not require attributing the change in the locomotion behaviour to brain-initiated mechanisms.

Rats rendered akinetic by dopamine depletion produced by a neurotoxin (6-OHDA) can be induced to walk with injections of atropine, an antagonist for the neurotransmitter, acetylcholine. The rat begins in its typically hunched (normal) posture, but once it begins walking, with every step it takes forward, its body becomes more and more elongated and although the rat begins walking with the lateral pattern, it switches to the diagonal pattern (Pellis et al., 1987). Detailed kinematic analyses revealed that the rat's steps with its forepaws had longer stride lengths than normal and those with its hind paws had shorter stride lengths than normal, so that, over several steps, the rat's body became elongated. By tracking its centre of mass, it became clear that once the rat had reached a certain degree of elongation, stepping with the lateral gait led to postural instability because its hind foot was too far from its ipsilateral forepaw (i.e., the one on the same side of the body), so that lifting that forepaw would lead the rat to fall over. It was at this point that the rat switched to stepping forward with its contralateral forepaw (i.e., the one on the opposite side of the body). The most parsimonious explanation for the switch in the rat's gait is that changes in the configuration of its body and the changes in postural stability that arose from that bodily; change, forced the rat to adopt a gait pattern that enabled it to remain standing. That is, the configuration of the rat's body explains its switch in gait; it need not involve a drug-induced activation of brainstem neural circuits.

Comparative studies of rodents with long bodies and short legs similarly show that, in some situations, such as when walking backwards up a slope, walking with the lateral gait can lead to postural instability, which can then lead to switching gaits (Eilam, Adijes & Velinsky, 1995). The same applies to some rodents in their early phases of development as their body morphology changes (Eilam, 1997). Thus, while this chapter and Chapter 2 describe two different ways in which brain mechanisms can contribute to the overt expression of behaviour (by controlling perceptions and by intrinsic motor organisation), the given examples illustrate that patterns in behavioural expression may sometimes arise not from the brain but as a consequence of the peripheral anatomy or in the organisation of the surrounding environmental context. These non-brain-derived regulatory factors are the topic of Chapter 4.

4   The Brain Is Not Alone

> [T]he multicausality of action, including the purely physical, energetic, and
> physiological components traditionally thought to be psychologically uninteresting
> but...(are) now recognized to be essential in the final movement patterns produced.
>
> Thelen (1995, p. 81)

A five-day old rat pup is most likely to engage in a 'corkscrew' action when
attempting to right itself (see Figure 1.3). As noted earlier, this corkscrew
action arises from the co-activation of tactile-induced righting by the
forelimbs and hindlimbs. Because, at that age, the pup's paws are immature, its ability to gain traction on the ground is limited and so its loss of
paw contact with the ground can lead to it extending its limbs in opposite
directions, and so producing the back-and-forth corkscrew-like bodily
configuration (see Figures 1.3 and 1.5b). This leads to righting that is
much delayed or can even lead to the animal failing to right itself, instead
lying on its back exhausted after several minutes of failed attempts to gain
traction with the ground.

We never observed such extended, unproductive righting when young
rats were in their home nests, with their siblings and mother. We suspected
that the presence of nest debris, siblings and the mother could be irregularities in the environment that could provide pups with a purchase for
gripping or pushing. Therefore, we manipulated the substrate on which
we tested them, by making it smooth or rough, which changed how easy it
was for them to gain a grip. This had an effect: the rough substrate was
easier for the pups to grip and led to them righting faster with a truncated
corkscrew, whereas the smooth substrate led to a greater delay in their
righting and a more exaggerated corkscrew (V. Pellis, Pellis & Teitelbaum,
1991). Similarly, when comparing the early onset of righting across different species, there were variations that drew our attention to another
important factor shaping the overt behaviour of the animals – their body

morphology. At around five days of age, the forelimbs and the hindlimbs of rats are not that different in size; however, at an equivalent age, the forelimbs of Syrian golden hamsters are markedly more developed than their hindlimbs and the hindlimbs of Mongolian gerbils are markedly more developed than their forelimbs. For these two species, while, like rats, they begin righting with co-activation of fore and hindquarter righting, in the hamsters, their forepaws are more successful in gaining the advantage and righting first, and in the gerbils, their hindpaws are more successful in gaining the advantage and righting first (unpublished observations). That is, while all three species go through a corkscrew stage in the development of righting, their different body morphologies influence the strength of the antagonistic forequarter and hindquarter movements and so how long righting is delayed.

What the example of the corkscrew pattern of righting tells us is that, as well as age-related changes in the neural mechanisms that control righting, the context in which righting is tested and differences in body morphology also need to be taken into account. This means that sometimes the overt behaviour that is measured may not be a true reflection of the neural mechanisms, but instead, may be a by-product of the testing environment and/or the subject's body morphology (Barrett, 2011; Clark, 1996). This chapter will use some examples to illustrate how the contribution of context and morphology can be disassociated from brain-related influences.

Environmental Context

Measuring the path travelled is a common way to assess how animals move about in their surroundings. Measuring can sometimes be automated, producing a detailed map of movements in space, either in the wild (e.g., Tomkiewicz, Fuller, Kie & Bates, 2010) or in the laboratory (e.g., Gomez-Marin, Partoune, Stephens & Louis, 2012). In the laboratory, the environment can be controlled, reducing the impact of the surroundings on the trajectory taken by the animal. For example, in a circular arena, with smooth walls, the initial path travelled by a mouse is also smooth and circular as it keeps its vibrissae in contact with the wall. As the mouse becomes more familiar with its surroundings, it makes excursions away from the wall, thus interrupting this smooth trajectory (Golani, 2012). The path travelled would become seemingly more erratic if the mouse were observed in a real-life setting, in which the regularity of the surroundings were not prescribed by the experimenter. In such a case, the irregular projections of vertical surfaces would lead to an irregular trajectory of movement as the animal maintains vibrissae contact with the surface. Given that it is well known that small mammals – such as mice and rats – are highly thigmotactic, researchers would be unlikely to interpret every deviation in the path travelled as some brain-generated process. The mapping of the shape of the vertical surfaces would account for the shape of the trajectory of movement by the animal.

Of course, as already noted, animals are not limited to following a simple rule, such as keeping in contact with a vertical surface when exploring; rather, they gradually build a map of the space by establishing home bases and adjusting their movements to and from such locations, enabling them to venture into the open areas of the arena (Golani, 2012). Similar patterns are present when rats are in situations in which they need to rely on visual cues (e.g., Vorhees & Williams, 2006) or proprioceptive and vestibular cues (Wallace, Hines, Pellis & Whishaw, 2002). That a map is formed which allows the animal to return to a safe sanctuary is reflected in the animals following a straight-line path at a higher speed when moving towards home compared to when they are moving away from home (Drai & Golani, 2001; Whishaw, Cassel & Jarrard, 1995). For example, during the third week after birth, young rats use their littermates as home bases from which they make excursions into the open field. From the outset, the homeward path, back to the huddled littermates, is more direct than the outward path (Figure 4.1). Many exploratory movements can be accounted for by the sensory cues in the environment. Again, for researchers working on exploratory behaviour, this is a given that is taken into account when interpreting the movements measured. It is not such a given in many other behavioural contexts, such as in animal combat.

In many studies of animal combat, predefined behaviour patterns thought to be typical of species are scored, so that a numerical value of their occurrence can be obtained, and where possible, the probability of transition from one behaviour pattern to another can be estimated (e.g., Breed, Meaney, Deuth & Bell, 1981; Guerra & Mason, 2005). The numerical differences are typically thought to reflect species-typical preferences in fighting tactics (e.g., Colvin, 1973; Dempster & Perrin, 1989). But in such

Outward   Homeward

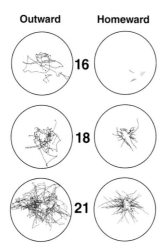

Figure 4.1 The path travelled in an open space is shown for rats over the third week post-birth (days 16–21) under infrared lighting (limiting visual guidance). Note the trips away from the home base are longer and more complex (left) than the homeward trips (right). Reprinted from Loewen, Wallace and Whishaw (2005) with permission (Copyright © 2005 John Wiley & Sons, Inc.)

analyses, simpler reasons for species differences may be overlooked. As already discussed, a simple way to induce fighting in rats is to place an unfamiliar male in the home cage of a resident, allowing the resident-intruder combat to be videotaped and analysed (Fernández-Espejo & Mir, 1990). However, the so-called preferred tactics may differ with the size of the test enclosure. In more constrained enclosures, the attacked intruder is more likely to roll over onto its back to protect its rump, whereas in a larger enclosure, the intruder is more likely to flee from the oncoming resident (Adams & Boice, 1989). In square or rectangular test enclosures, there are corners that enable the intruder to press its back against while standing upright and facing its attacker. Where corners are readily available, more upright defence manoeuvres may be scored than when the intruder is attacked in the open (Pellis, Pellis, Manning & Dewsbury, 1992). The environmental context can therefore have a causal influence on the tactics adopted and, so, on the behaviour patterns scored for numerical comparison. One way to overcome this environmental factor is to be selective in the context scored. An example will illustrate the logic of this procedure.

When using the resident–intruder paradigm to compare the combat of two species of voles – prairie and montane voles – we found that, in the former, residents were more likely to chase intruders, whereas, in the latter, residents were more likely to use the lateral attack manoeuvre on intruders. Correspondingly, in the former, intruders were more likely to flee from residents, whereas, in the latter, intruders were more likely to stand upright and face residents (Pierce, Pellis, Dewsbury & Pellis, 1991). As has already been discussed, for these and related rodents, the target of attack is the rump and lower flanks (Pellis, 1997) and the tactics of attack are highly correlated with the tactics of defence (Blanchard, Blanchard, Takahashi, Kelley, 1977; Pellis & Pellis, 1992). To determine whether these species differences reflect species-typical differences in attack or defence, a strategy to factor out the influence of the environment had to be employed. Only attacks when the intruder was in the middle of the enclosure were scored, thus eliminating the potential confounding influences of walls, especially corners. Of attacks in the open, only those that were from the rear were scored, thus eliminating the potential confounding influences of the variable orientation of the attacker. Intruders of both species were thus compared in exactly the same configuration, and the immediate defensive response when the attacking resident lunged at the intruder's exposed rump was scored. Once the resident pounced, thus releasing its support on the ground, and immediately before the contact on the intruder's rump with its teeth was made, the attacking resident could no longer influence the defensive tactic initiated by the defending intruder. A context was therefore selected that minimised the ability of one animal to influence the action of the other. In this way, at least for a brief moment, the preferred action by one animal, the defending intruder in this case, could be assessed. We term such moments 'decision points', because they give insight into the animal's preferred tactics (Pellis, 1989).

Using this set of criteria to compare preferred differences in defence between the intruders of the two species revealed that there were species differences (Pellis, Pellis, Pierce & Dewsbury, 1992). Montane voles were more likely to turn to face their attacker and did so by adopting an upright posture, whereas prairie voles were more likely to flee (Figure 4.2). Comparing the cases in which intruders from both species adopted the upright facing defence to block the resident's attack on their rumps revealed that, while montane voles were more likely to maintain that

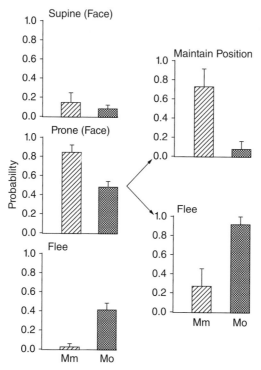

Figure 4.2 The behaviour of the intruders (i.e., defenders) of both species is compared in two behavioural contexts: On the left, the preferred tactic to defend the rump from a bite from the rear, and on the right the preferred tactic from the upright facing position is shown. In both cases, prairie voles (Mo) were more likely to flee, whereas montane voles (Mm), in both cases, were more likely to face their attacker. Reprinted from Pellis, Pellis, Pierce and Dewsbury (1992) with permission (Copyright © 1992 John Wiley & Sons, Inc.)

posture, prairie voles were more likely to turn and flee (Figure 4.2). Does this reflect a persistent species difference in preference for fleeing? The more likely explanation comes from differences between the two species in the behaviour of the attackers. The prairie vole resident was more likely to lunge upwards and direct a bite to the upright intruder's face, the body area targeted during high-intensity aggression (Pellis, 1997), whereas the montane vole resident was more likely to continue to manoeuvre to try and bite the intruder's rump. That is, remaining in the upright defence posture was not an effective strategy for continued defence by prairie voles as it placed their faces at risk of being bitten. As the trial progressed and prairie vole intruders experienced such attacks to their faces, they became increasingly less likely to remain in a fixed defensive posture (Pellis, Pellis, Pierce & Dewsbury, 1992).

By taking into account the confounding effects of the environment, a species difference in the preferred defensive tactic was identified and this preference was further found to reflect a species difference in the offensive behaviour of the residents. These studies revealed something else – neither species was likely to adopt a supine defence tactic to extricate their rumps from imminent attack. This species commonality reveals the importance of another, non-brain-dependent factor – body morphology.

Body Morphology

Even though most mouse-like rodents use the supine defence as a tactic to avoid bites to their rumps and lower flanks, there is variation in the frequency of its use. Further, even when scored in the decision point characterised above, thus removing contextual confounds, some species use this tactic more often than others (e.g., Pellis & Pellis, 1988b, 1989, 1992; Pellis, Pellis, Manning & Dewsbury, 1992; Pellis, Pellis, Pierce & Dewsbury, 1992). While these differences could reflect brain-based biases, these rodents also vary in body morphology, and this needs to be taken into account before it is determined that there are differences in brain mechanisms. Therefore, six rodent species with marked differences in body shape were compared (Pellis, 1997). Two species, Syrian and Djungarian hamsters, have short, squat bodies with relatively short limbs; two others, montane and prairie voles, have long bodies with short legs; and, finally, two other species, house mice and grasshopper mice, have a body shape and limb lengths intermediate between the voles and the hamsters. A ratio of head–body length divided by body weight was used to give a numerical index for these body differences, so that the value for the Syrian golden hamster was the smallest and that for the prairie vole the largest. By using the above-defined decision point to evaluate the percentage of defensive responses involving supine defence, there was a significant correlation between the

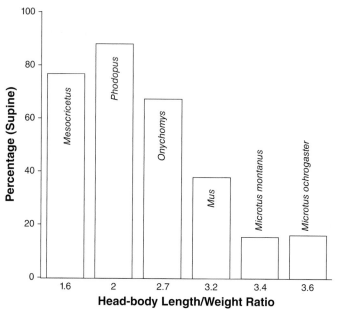

Figure 4.3 The probability of rotating to supine when being bitten on the rump is contrasted in the length-to-weight ratio of six species of rodents. The two hamster species, the Syrian golden hamster (*Mesocricetus auratus*) and the Djungarian hamster (*Phodopus campbelli*), with the smallest ratio, have the greatest likelihood of rotating to supine, whereas the two vole species with the largest ratio, the montane vole (*Microtus montanus*) and the prairie vole (*M. ochrogaster*), are the least likely to rotate to supine. The two other species, house mouse (*Mus domesticus*) and the grasshopper mouse (*Onychomys leucogaster*), are intermediate for both measures. There was a significant negative correlation between the two values ($r_s = -0.89$, $p < 0.05$). Reprinted from Pellis (1997) with permission from John Wiley & Sons (Copyright © 1997 Wiley-Liss, Inc.)

two variables, with animals that have a smaller body-weight ratio employing the supine tactic more often (Figure 4.3).

Detailed analyses of the movements performed during the execution of these tactics provide clues as to why animals of differing body morphology may preferentially use the supine tactic, whereas others may be less inclined to do so. In the context analysed, the defending animal used the supine defence to extricate its rump from an imminent bite on its rear by its opponent. To do this involves cephalocaudal recruitment of the body, beginning with the head. Given that, in long-bodied animals, the rump is further from the head than in short-bodied animals, it is likely that the recruitment of the pelvis to rotate to supine takes a longer time in them. If this is true, then rotating a short body should extricate the rump from harm's way faster than a long body. Comparing the duration of rotation between the two species with the most different body ratio (Figure 4.3) showed that Syrian golden hamsters achieved the fully supine position nearly twenty percent faster than prairie voles (Pellis, 1997). Given the imminence of the bite, gaining a hundred millisecond advantage in speed of rotation can make all the difference.

Similarly, comparing the movements used to deflect a bite to the rump – using turning to face the attacking animal in a prone position, which leads to the upright posture – showed a disparity in effectiveness based on body morphology. The defending long-bodied voles bent their bodies laterally in the horizontal plane, so that their heads, necks, and upper bodies rotated to face their opponent's head. This manoeuvre could be coupled with a lunge to bite the attacker's face. To avoid being bitten on the face, the attacker moved backwards. This gave the defender the opportunity to rotate its hindquarters to be aligned with its forequarters and rear upright (Figure 4.4a). In contrast, short-bodied hamsters began by turning their head and neck laterally and then to rear onto their hindfeet as they rotated their upper body around their longitudinal axis. The defending hamster then rotated its hindquarters, moving its rump away from its attacker as its mouth lunged downward onto its attacker's head (Figure 4.4b). Again, small differences can be decisive: for voles, the very first phase of the turn produced a defensive bite, whereas in hamsters, the defensive bite occurred in the later phases of the sequence of body movements. Quite simply, there are more body movements for hamsters that precede the retaliatory bite. The apparent advantage in supine defence for short-bodied animals and the apparent advantage for upright defence in long-bodied animals may account for the species differences in how frequently they use these tactics (Pellis, Pellis, Pierce & Dewsbury, 1992; Pierce, Pellis, Dewsbury & Pellis, 1991).

Avoiding Assumptions and Testing Hypotheses

After a long dominance of 'it all emanates from the brain' approach to observed behavioural output, the importance of context and body

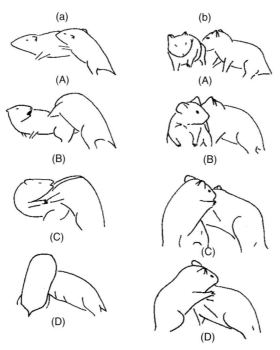

Figure 4.4 By turning laterally toward an attacker, the rump defense is shown for two species of rodents. a: The attacking prairie vole (on the right) lunges and bites the defender's lower flank (A). The defender turns its head and neck laterally toward its attacker, (B) and then, with a lateral turn of its upper body, the defender lunges at the side of the attacker's face (C). Once the defender delivers a retaliatory bite, it pivots around its hindlimbs and, withdrawing its lower body, rises to an upright posture while simultaneously pressing its counterattack (D). b: The attacking Syrian golden hamster (on the right) lunges and bites the defender's rump (A). The defender turns its head and neck laterally toward its attacker (B), but then, to complete its turn towards the attacker, the defender raises its forequarters while rotating around its longitudinal axis (C). Once the defender faces the attacker, it pivots around its hind limbs, withdraws its lower body, and presses the counterattack (D). Reprinted from Pellis (1997) with permission (Copyright © 1997 John Wiley & Sons, Inc.)

morphology in shaping the form of that behavioural output has come to the fore (Barrett, 2011; Clark, 1996; Thelen, 1995). However, the pendulum has swung so far in the other direction that it is often assumed that species, age or sex differences in body morphology account for differences in overt behaviour. Structure arising from outside the brain has become the default explanation: 'the brain does not explain this behavioural pattern unless you show otherwise'. For example, aye-ayes are nocturnal, tree-living lemurs from Madagascar that have exceptionally long digits which are used to fish beetle grubs out of hiding from under tree bark. Their thumbs, relative to the other digits, are short. In captivity, they are fed a variety of food items, including large ones such as slices of melon and whole eggs. Sitting in the dark, in their cage, you can hear them collecting such food, but you can also hear the splat of food falling to the ground. An obvious assumption is that their long, specialised fingers and short thumbs compromise their ability to hold such difficult to grasp food items. However, a detailed examination of the food handling of aye-ayes compared to other lemur species with less specialised digits revealed that aye-ayes have, in fact, more sophisticated grasping and thumb use than the others. It is important to note, however, that morphological changes in the hands may be important for facilitating these novel patterns of grasping (Hartstone-Rose, Dickinson, Boettcher & Herrel, 2019). Whatever the inter-play between hand morphology and grasping pattern, the reason for all the dropping of food is that, typically, aye-ayes pick up a food item and then carry it upwards, which requires holding the food in one hand while climbing with the other three limbs – and accidents do happen. Typically, other species of lemurs eat the food where they collect it – no climbing, no accidents. Indeed, one of the aye-ayes that we observed habitually ate food on the ground where she found it, and she never dropped a single item (Pellis & Pellis, 2012).

As with all factors influencing the production of overt behaviour that we have discussed so far, the role of context or body morphology in any given case is a hypothesis that needs to be tested, not assumed. For example, when rats eat a portable piece of food, they pick the item up with their mouth, then grasp the item in their paws and manipulate it as they chew pieces (Whishaw, Dringenberg & Pellis, 1992). But being a social species, one rat eating food attracts the attention of its neighbours. The odour from an individual's mouth can provide information about the local availability of food and this can be obtained from sniffing and sampling what another rat has eaten or is eating (Galef, 1996). Consequently, a rat eating a piece of food not only attracts the interest of a neighbour, but that neighbour may also attempt to take that food item away. To protect the food item held in its mouth, the owner pivots away from the potential robber (Figure 4.5).

(a)

(b)

(c)

(d)

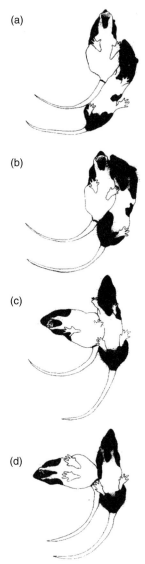

Figure 4.5 Food robbing and dodging are shown in rats. The robber, on the right, approaches the mouth of the rat that is eating. This leads to the rat that has been eating swerving laterally away from the robber. Reprinted from Whishaw (1988) with permission (Copyright © 1988 Elsevier, Ltd.)

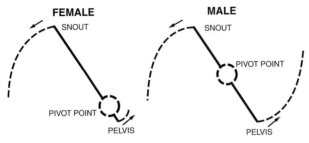

Figure 4.6 Schematic drawings show the trajectories of the snout and the pelvis during dodging by a female and male rat. Note that the pivot point in the female is closer to the end of its pelvis, whereas in the male, it is more centrally located. Because of this, males make a larger excursion of the pelvis. Reprinted from Pellis, Field, Smith and Pellis (1997) with permission (Copyright © 1997 Elsevier Ltd.)

This pivoting, or dodging, moves the owner's mouth and their food item away from the other animal, the robber (Whishaw, 1988).

Using this 'robbing and dodging' paradigm to compare food protection in male and female defenders led to an unexpected finding: the sexes both dodged, but how they achieved the dodge was different (Field, Whishaw & Pellis, 1996). The females pivoted around their pelvis, whereas the males pivoted around their mid-body (Figure 4.6). To achieve this, the sexes had different patterns of stepping. From our first observations, the response of reviewers was, 'well, of course, females and males are built differently'. This led our then graduate student, Evelyn Field, on a long journey to determine whether indeed the sex difference was a by-product of differences in body morphology. First, the sex difference was found to be comparable across a variety of motor actions (e.g., standing and turning, reaching for a food pellet, bracing, righting). In each case, the integration of fore- and hind-quarter movements was different in males and females. Second, these differences were present even when the sexes were matched for size. Third, these differences were present even before puberty, before sex differences in the morphology of the pelvis emerge. Fourth, these differences were present even in males with a genetic mutation that led to them having a masculinised brain but a feminised body. Fifth, these differences were abolished in rats with the same sex-typical body, but suffering from a particular type of brain damage. In order to prove our case, we had to go above and beyond to demonstrate that the consistent sex difference in movement organisation was a result of brain differences and not to differences in body morphology (Field & Pellis, 2008; Field & Whishaw, 2008).

The bottom line is that positing that differences in behaviour across species, sexes or ages arise from the influences of context or body morphology is a hypothesis, and like any hypothesis, needs to be put to the test. No assumed explanation is privileged, be it perceptual constancy (i.e., a perceptual bias), intrinsic behavioural organisation (i.e., a motor bias) or context and body morphology (i.e., biases external to the brain). All need to be applied and tested for their capacity to account for the overt behaviour being observed. To see how all these principles and their testing can be applied to unravelling the organisation of a complex behaviour and then how to use that understanding to develop appropriate behavioural markers suitable for quantitative analyses will be detailed in Chapter 5.

5 Bringing It All Together: Steps in the Descriptive Process

Description is never, can never be, random; it is in fact highly selective, and selection is made in reference to the problems, hypotheses and methods the investigator has in mind... However, if we overdo this in itself justifiable tendency of making description subject to our analytical aims, we may well fall into the trap some branches of Psychology have fallen into, and fail to describe any behaviour that seems 'trivial' to us; we might forget that naïve, unsophisticated, or intuitively guided observation may open our eyes to new problems.

Tinbergen (1963, p. 412)

The young of many mammals and birds engage in vigorous peer-to-peer play (Burghardt, 2005; Fagen, 1981). Such play is most typically in the form of rough-and-tumble play or play fighting (Aldis, 1975). While play fighting resembles real fighting, it has features that make it playful and so distinctive (Pellis & Pellis, 2017; Smith, 1997). Juvenile rats are prodigious players, making them prime candidates for laboratory studies of this behaviour (Pellis & Pellis, 2009; Siviy, 2016; Vanderschuren, Achterberg & Trezza, 2016). Typically, rats chase after one another, with one pouncing on the other; this may lead to a playful wrestle lasting three to four seconds before they move apart and perhaps resume chasing one another. As can be seen in Figure 5.1, wrestling involves the animals grappling with one another, and by doing so, they undergo many changes to their postural configurations. But playful interactions are highly variable, with some pounces not leading to chasing or wrestling. Some wrestles begin when the animals are in close proximity and so do not begin with a pounce. There are too many details and variations to score every facet of this behaviour, so the problem has been that of reducing this complexity to some relatively simple behavioural markers that can be scored reliably.

The regularity of the jumps, runs, pounces, wrestles, and particular postural configurations during this form of play led to two main ways in which to quantify this behaviour in rats. One approach has been to count all

Figure 5.1 Two juvenile male rats, at about 35 days old, are shown engaging in a play fight in which they compete for access to the nape of each other's necks. Reprinted from Pellis and Pellis (1987) with permission (Copyright © 1987 John Wiley & Sons, Inc.)

the different, readily recognizable, behavioural elements – such as pounces, chases, and wrestles – and then add them up to obtain an overall score of how much the rats play per unit time. Another focusses on a more limited subset of postural configurations that yield the highest inter-observer reliability and are correlated with overall rates of playing. Even though different laboratories use idiosyncratic versions of these approaches (Blake & McCoy, 2015), on the whole, all the variants of these measuring schemes provide relatively accurate assessments of how much the rats play. However, these descriptive measures, whether explicitly stated or not, are not divorced from the presumed underlying organisation of play behaviour. Rats are motivated to play, with play – no matter how simple or complex – being viewed as a unitary behaviour. Therefore, irrespective of the scheme used to measure play, all changes in frequency can be interpreted as equivalent ways of assessing changes in the mechanisms that regulate play behaviour. But what if this assumption is incorrect? If play fighting in rats involves more than one causal process in producing the overt behaviour measured, then using these numerical approaches can make it difficult to assign changes in the amount of play to particular causes (Himmler, Pellis & Pellis, 2013).

In this chapter, we will use the play fighting of rats to illustrate three steps in the descriptive process. First, we will show how the principles discussed in the preceding chapters can be applied to gain insight into the organisation of behaviour. Second, we will demonstrate how useful behavioural markers can be abstracted so that they can be scored quantitatively in a way that reflects that underlying organisation. Third, by explicitly linking the abstracted behavioural markers to the organisation of the behaviour, these abstracted markers can be tested for their validity in representing that organisation. This iterative process can reveal novel aspects of organisation that then require the abstraction of new markers to assess, quantitatively, that newly revealed organisation. Description and the measurement of specific behavioural markers is thus an evolving process.

Before delving into the details of play fighting in rats, it may be worthwhile to reiterate the four principles detailed in the preceding chapters. Two are based on brain-derived mechanisms and two on processes that are independent of the brain, but they do have a two-way interaction (Table 5.1). The perceptions that are salient and are controlled by the animal are brain-based, but their availability depends on properties of the surrounding environment, which are independent of the animal's perceptual preferences, and so are non-brain-based. This distinction and commonality is important because the animal likely controls many perceptions (Powers, 2005) and that which controls the animal's behaviour will depend on which perceptions are available in a given context. Similarly, how the available

Table 5.1   The two-way division of the four principles of behavioural organisation

Principles of organisation	Externally derived information	Internally dependent constraints
Brain-based	Controlled perceptions	Intrinsic motor bias
Non-brain-based	Environmental context	Body morphology

motor actions are used depend on how they are coordinated together in the neural circuits in the nervous system, and so are brain-based. In turn, how those actions are executed and which options may have priority is, in part, dependent on non-brain-based biomechanical constraints imposed by the animal's body morphology (Pellis, Pellis & Iwaniuk, 2014). Consequently, the animal's body can influence behavioural output in two ways: via brain anatomy and physiology and via the anatomy and physiology of the rest of its body.

While for any given behaviour these four principles may interact, it is useful to focus on each independently by eliminating the influence of the others. Then, once the role of each is understood, how they may interact becomes easier to identify and characterise. Therefore, in what follows, play fighting will be viewed from the perspective of each principle.

The Perceptual Constants of Play Fighting

A closer inspection of Figure 5.1 reveals that the locus of the action is the interplay between the two rats' snouts and napes. The rat on the left approaches the second rat (a), and then, from its rear, reaches towards its opponent's nape with its snout (b). However, before contact can be made, the rat rotates around its longitudinal axis (c) to face the approaching attacker (d). By moving forward, the attacker pushes its opponent onto its side (e). The rat on its side then rolls over onto its back as the other rat continues to reach for its nape (f–h). The rat lying on its back then reaches towards its partner's nape (i), but this fails due to its partner using its hind foot to push its opponent's head away (j, k). The rat on top (l) is eventually pushed off by the supine animal (m), which then regains its footing (n) and lunges towards its partner's nape (o). When the recipient of a playful

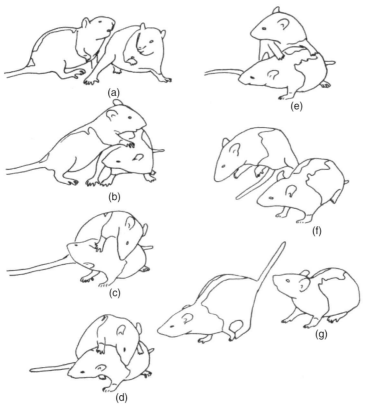

Figure 5.2 Two juvenile male rats, at about 35 days old, are shown engaging in a simple play fight. The animal on the right is shown approaching another rat, then contacting it on the nape and leaping away. The recipient of the attack does not defend itself. Reprinted from Pellis (1988) with permission (Copyright © 1988 John Wiley & Sons, Inc.)

approach to the nape does not respond, the importance of the nape becomes strikingly evident (Figure 5.2). The rat on the left approaches its partner (a), reaches up towards its nape (b), and then rubs its snout into the nape area (c and d). Once it has finished its nosing, the rat raises its head (e) and leaps away (f and g).

The two processes that emerge from what has been described above are that: (i) a pair of rats will compete to gain access to each other's napes (Pellis & Pellis, 1987) and (ii) a successful contact leads to rubbing a partner's nape with one's snout (Pellis, 1988). Indeed, if either the snouts or the napes of the interacting rats are anaesthetised, the typical pattern of play is blocked (Siviy & Panksepp, 1987). The inescapable conclusion is that play fighting in rats involves attack and defence of the nape. Closing the distance between the tip of your snout with your partner's nape is one controlled variable (CV), while keeping the distance between your nape and your partner's snout above zero is the other controlled variable (see Chapter 2). As the two perceptions are in conflict, the manoeuvres by one rat create a disturbance to the CV of the other rat, and, in turn, its compensatory manoeuvres create disturbances to the CV of the first rat. Thus, the movements of the two animals are correlated: the actions of one rat cannot be understood without taking into account the actions of the second rat (Pellis & Bell, 2020; Pellis & Pellis, 2015). This correlation of actions creates problems of interpretation for some commonly used measures.

During playful wrestling, a common configuration for a pair is for one partner to lie on its back while its partner is standing on top (see panel h in Figure 5.1). This pinning configuration is not only highly correlated with play fighting overall, but it is also readily recognisable even when scoring play in real time (Panksepp, 1981), and so has been widely used as a measure of play frequency (Blake & McCoy, 2015; Himmler, Pellis & Pellis, 2013). However, when placing this configuration in context, it can be seen that the rat is rolling over onto its back as a defensive action to protect its nape from its partner's attack (Figure 5.1). Therefore, as the majority of pins arise from such a defensive action (Pellis & Pellis, 1987), a change in the frequency of scoring pins may arise from three, distinct causes – the frequency of nape attacks may change, the likelihood of defending the nape may change, or the likelihood of using rolling over to supine as a defensive tactic may change. As a consequence, an experiment that yields a reduced frequency in pinning cannot unambiguously be interpreted as a treatment that leads to a reduced motivation to engage in play. For example, a juvenile rat that has had its cortex ablated shortly after birth initiates just as many nape attacks and is as likely to defend its nape against such attacks as its intact siblings. However, while for the intact siblings the supine defence tactic is the modal tactic used, for the rat with its cortex ablated, other tactics, involving standing, are used more frequently (Pellis, Pellis & Whishaw, 1992). These alternative defensive tactics are less likely to lead the pair to adopt the pin configuration. Thus, in this case, a reduced frequency of pinning is not a reflection of a reduced motivation to play, but rather, is a reflection of a change in how the rats defend themselves during play. Simply scoring pinning rather than the range of tactics used to defend the nape does not reveal how play is changed by the experimental manipulation (Panksepp, Normansell, Cox & Siviy, 1994; Pellis, Pellis & Whishaw, 1992).

An attempt to factor out the role of attack in producing pins was to introduce a measure for contacting the nape (Panksepp, 1998). To qualify, an attack has to achieve a highly specific configuration. The attacking rat has to have both of its forepaws on its partner's back (usually one paw on top of the other's head and one on its shoulders) while its snout is in contact with its partner's nape (see panel d in Figure 5.2). Again, the advantage of scoring such a highly specific configuration, variously labelled 'nape contact' or 'nape pounce' (see Blake & McCoy, 2015, for a review of the literature), is that it is easily recognisable, it can be readily scored in real time and it has a high inter-observer reliability. However, for reasons that we will explore in greater detail, this measure is also problematic as it again confounds the attack and defence behaviour of the two partners. In effect, a nape contact reflects a successful nape attack (as in Figure 5.2), whereas many nape attacks may be launched without achieving such a configuration (as in Figure 5.1). Consequently, nape contacts are highly likely to underestimate the frequency of playful attacks because the defensive behaviour of the partner can prevent the successful achievement of the highly configured contact with the nape. Therefore, both pins and nape contacts confound the combined actions of the two partners, so that neither is a pure measure of either attack or defence.

When applied to play fighting in rats, the principle that behaviour is used to gain or maintain particular perceptions (Chapter 2) shows that the nape of the neck is the target of such an encounter. As illustrated by the limitations in scoring 'pinning' and 'nape contact', abstracting behavioural markers in a way that does not take the dynamic competition between the partners into account can be misleading.

Play Fighting as a Behaviour System

Because play in rats involves attack and defence, many of the actions and postural configurations are similar to those seen in the aggressive interactions of adults. Consequently, some researchers assumed that play fighting in young rats is an immature, incomplete version of serious fighting (e.g., Hurst et al., 1996; Silverman, 1978; Taylor, 1980), just like the corkscrew is a reflection of immature righting behaviour (see Chapter 1). Although this may seem like a fair assessment of 'play fighting', re-examining play fighting across a wide diversity of species from the perspective of the advantage competed over during such play yields a different picture. Although, in many species, the advantage sought in play fighting is biting or striking the same body targets as in serious fighting in adults (Aldis, 1975), this is not the case universally. For some species, the targets competed over are the same as those in predation, sex and social grooming (Pellis, 1988; Pellis & Pellis, 2017) and some will engage in multiple forms of play fighting (Pellis & Pellis, 2018). So why do rats compete to rub each other's napes with their snouts? It is not what they do in serious fighting, where bites are directed at lower dorsum and flanks or the face (Blanchard & Blanchard, 1994; Pellis, 1997), but it is what an adult male rat does as a prelude to mounting a receptive female during sexual encounters (Whishaw & Kolb, 1985).

A comparison across a diverse range of rat-like rodents showed that many engage in play fighting that simulates pre-copulatory behaviour. Some rodents are like rats and nose the nape (e.g., deer mice), and others target completely different body areas, such as licking and nuzzling the mouth (e.g., Djungarian hamsters) (Pellis & Pellis, 2009). In addition, some the of the runs and jumps present in the peer–peer play of rats further points to play being associated with sex. During adult sexual encounters, receptive females dart and hop past or near the male, which appear to be solicitation behaviours (Thor & Flannelly, 1978). These darts and hops are also present during peer-to-peer play, in between bouts of play fighting (Thor & Holloway, 1986). Juvenile Syrian hamsters compete to nibble the cheeks of their partners during play fighting (Pellis & Pellis, 1988a), which is what male Syrian hamsters do to receptive females during pre-copulatory encounters (Pellis & Pellis, 1988b). However, unlike rats, there are no darts, hops, or runs during sexual encounters (Carter, 1985; Siegel, 1985), and these are also absent during juvenile play (Pellis & Pellis, 1988a). That is, the species differences in play match the differences in adult sexual behaviour. Given that nuzzling the nape and the associated behaviour patterns present in the play of rats are the same suite of behaviours that are correlated with one another during sex, it can be concluded that it is the sex behaviour system that is activated during play (see Chapter 3). For rats, then, like many other rodents, play fighting mostly simulates sexual interactions, not serious fighting. This has two important implications for what to measure.

First, the behaviours associated with play fighting have their organisational roots in sex, not aggression. Consequently, the actions used to gain access to or defend the target are ones that are more closely aligned to sex. This is strikingly illustrated by two, closely related species of voles. In both species, the target of competition in both sex and play is the nape, which is nuzzled if contacted. However, the tactic most frequently used to defend the nape differs between the two species, both in sex and play. In one species, the supine defence is used more frequently, whereas in the other, the standing upright defence is used more frequently (Pellis & Pellis, 1998). That is, both the similarity in the target and the difference in the tactics used during play between these two species can be accounted for by the species-typical organisation of sexual behaviour. In rats, while the tactics used in play are the same as those in sex, the tactics used most often in sexual encounters are not the ones used most often during play fighting (Pellis & Iwaniuk, 2004). The differences in the use of tactics in voles and rats has further implications as to what constitutes suitable behavioural markers to score for play, a topic to which we will return later in this chapter. For now, the important lesson to draw is that the differences in targets of attack and defence in play as compared to serious fighting provide strong evidence that play fighting in rats is not immature aggression.

Second, even though there are gross similarities in the tactics used in the playful and serious fighting of rats, the different targets have an important impact on the structure and function of those tactics. For example, in both serious fighting and play fighting, the 'lateral posture' can occur. As explained in Chapter 2 (see Figure 2.5), during serious fighting, the lateral posture is used as a tactic of attack. Because of the risk of a retaliatory bite by the defender, adopting the lateral configuration enables the attacker to close the distance between its body and its opponent's while keeping its own head out of striking range. In play fighting, in which both partners typically attack and defend the nape (see Figure 5.1), such a protective component is not needed. Consequently, during play fighting, rats do not use the lateral tactic for offence. However, in play fighting, rats do use a defensive version of the lateral manoeuvre. When contacted on its nape from the side, a defender may swerve its head and nape away from its attacker while simultaneously moving its lower flank towards its attacker, blocking its further movement towards its nape (Pellis & Pellis, 1987). Therefore, in serious fighting, the lateral tactic is used for offence, whereas in play fighting, it is used for defence. Hence, the presence of a superficially similar looking lateral tactic in both forms of fighting is not evidence that playful fighting is an immature version of aggression.

The organisation of play fighting in rats around attack and defence of the nape highlights another problem with scoring markers such as pinning. The pin configuration is the end point of several actions by both partners (Figure 5.1), and, as is shown in Chapter 3, end point measures can be misleading. A comparison of play fighting across five strains of rats revealed that the frequency of pinning varies markedly between them (Himmler, Stryjek, Modlińska et al., 2013; Himmler, Modlińska, Stryjek et al., 2014). Strain differences in the number of nape attacks launched were insufficient to account for these differences. More detailed analyses among the three most divergent strains, Long Evans, Sprague Dawley and wild rats, revealed how these differences arose (Himmler, Himmler, Stryjek et al., 2016).

The first important factor to note is how the rats end up in the supine position. When contacted on the nape, a defender can immediately roll over to the supine position, so extricating its nape from the reach of its partner's snout. This occurs as a continuous, fluid action, taking two to three video frames for cameras filming at thirty frames/second (Himmler, Pellis & Pellis, 2013). In other cases, the defender only partially rotates, so that while its head, neck and upper torso move away from its attacker's snout, its lower body is left prone, with its hind feet standing on the ground. Then, with the attacker's continued pushing and manoeuvering, the defender rotates, incrementally, until it is fully supine, but this can take six or more frames. In other cases, the defender stands and turns to face its attacker, and, as it does so, the attacker leaps forward, knocking the defender over onto its back.

In all three cases, the defender ends up being supine with the attacker standing on top, but how, in each case, the rats arrive at this point differs. The complete and rapid rotation to supine suggests that the defender preferentially engages in this tactic to protect its nape, whereas the supine

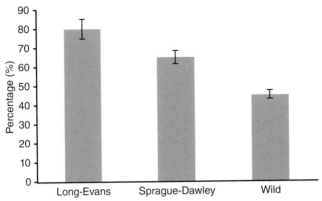

Figure 5.3 The mean and standard error are shown for the percentage of pins in three strains of rats (Long-Evans (LE), Sprague-Dawley (SD) and wild (WWCPS)) that arise from a defender actively rotating to the supine position. The alternative way to achieve the supine position is for the attacking rat to knock its partner over on to its back. Reprinted with permission from Himmler, Himmler, Stryjek et al. (2016)

position arising from the other two cases is imposed on the defender by its attacker. The percentage of pins achieved by the rat actively rolling over to supine revealed that rolling over to supine was used more often by the Long Evans rats (Figure 5.3). For some strains, therefore, being pushed onto their backs is more common than rolling over to supine.

To assess the preferred tactic of the defending rat, we use the first two to three video frames from the beginning of the manoeuvre, as this is prior to the confounding influence that results from the continuing attack manoeuvres of its partner (Himmler, Pellis & Pellis, 2013). Using this time frame to measure the strain differences revealed that the lower than expected magnitude of the difference in the pinning between the Long Evans and the Sprague Dawley rats could not be fully explained by the strain typical difference in preference for the supine defence. Again, the fact that the pin is an end point measure needs to be taken into account. While rolling over to supine was the first tactic used by the Long Evans rats, the Sprague Dawley rats attempted a variety of other tactics first and only resorted to rolling over to supine when the other tactics failed to extricate their nape from their attacker's snout (Himmler, Himmler, Stryjek et al., 2016). That is, the different preferences for attack and defence tactics account for some of the measurable differences across strains and these preferences arise from intrinsic, brain-derived differences (Himmler, Himmler, Pellis & Pellis, 2016). However, not all of the strain differences leading to overt differences in behavioural markers arise from such brain-level preferences.

Context and Morphology in Shaping Playful Actions

Features of the test enclosure influence how rats play. Rats engage in more play fighting in square or rectangular enclosures than in circular ones (Hole & Einon, 1984). Similarly, when we first started studying play fighting in rats, we tried different enclosures to maximise being able to view the rats' movements and discovered that context matters. For example, because the rats roll around during playful wrestling, we wanted to film the encounters from the side, from above and from below. But to film from below (for an illustration as to how this can be achieved see Field, Whishaw & Pellis, 1996), we could not use bedding, as this would obscure the view. Irrespective of whether tested in circular or rectangular enclosures, enclosures without bedding result in less play. In many interactions, one rat lunges and its partner turns to face it, but as it does so, the lunging rat makes contact, pushing its partner to the ground, resulting in an often audible thud. Bedding cushions the fall. The discomfort that comes from such falls is likely aversive in enclosures without bedding, which may reduce the playful interactions initiated and defended, making play fights less common. When more detail is needed, we insert a mirror on the back wall, so when filmed from the front, we can obtain a view of both flanks of the animals, so reducing the number of movements that are out of sight (Foroud & Pellis, 2003).

With regard to enclosure shape, recall that rats are thigmotactic (Chapter 2), so they tend to remain in close proximity to walls, with corners having the most attractive features as they enable greater body contact with vertical surfaces. Not unsurprisingly then, more play fights take place in the corners than elsewhere along the walls and more occur against walls than in the open area of the enclosure. Consequently, more play takes place in enclosures with corners than ones without. It is not only the amount of play that is affected, but also the actions that are performed during play.

The tactics used to extricate the nape from being contacted by a partner fall into two main categories: (1) *evasion*, in which the defender moves its

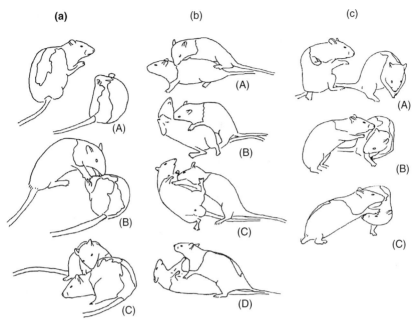

Figure 5.4 Three types of defense to a playful nape attack are shown in 61-day-old male rats (a–c). a. Evasion. Following the attacker's lunge at the nape (A, B), the defender swerves away from the attacker (C). b. Complete rotation. A nape contact from behind (A) leads the defender to rotate (B, C) until it is lying supine and is blocking its attacker with its outstretched paws (D). c. Partial rotation. A lunge to the nape from the side (A, B) is followed by a rotation of the head, neck and shoulders by the defender, withdrawing its nape from the attacker's snout (C). Reprinted from Pellis, Pellis and Whishaw (1992) with permission (Copyright © 1992 S. Karger AG, Basel)

nape away from its attacker by moving its head away. This can be achieved by dodging laterally (Figure 5.4a), running or leaping away and (2) *facing defence*, in which the defender moves its nape away by turning to face its attacker, thus juxtaposing its snout between its nape and its partner's snout. In turn, facing defence can be achieved in two, distinct ways. (i) The rat can rotate around the longitudinal axis of its body. This is usually associated with attacks from the side. (ii) The rat can rotate around its vertical axis, centred at the rump. This is usually associated with attacks from the rear (Pellis, Pellis & Dewsbury, 1989). Rotation around the longitudinal axis of the body begins with the head and progresses along the body, with the most extreme form leading the rat to roll into a fully supine position

(Figure 5.4b). Alternatively, the rat may stop rotating once it withdraws its nape from its partner's snout, so that its hindquarters remain prone, with either one or both hind feet maintaining contact with the ground (Figure 5.4c). As noted, when the rat adopts the complete rotation tactic for defence, the rotation is fast and continuous, and, once the rat is on its back, the options for further defence and counter-attack are relatively limited by the advantageous position of the rat on top, that is then able to use its forepaws to restrict the movements of its supine partner (see Figure 5.1). From the partially rotated position, the defender has more options.

By having removed its nape from the reach of its attacker's snout by partial rotation (Figure 5.4c, panel C), the defender can now push its attacker with its hind foot or by using a hip slam, or it can use its forepaws to grasp and push its attacker. Each of these options allows the defender to place the attacker off-balance, thus creating an opportunity for the defender to swerve laterally and flee. Alternatively, from the partially rotated position, the defender can rear up onto its hind feet while rotating its body so as to face its attacker fully, and, from this position, push the attacker over. This now increases the defender's ability to launch a counter-attack against its opponent's nape. Furthermore, from the horizontal position (Figure 5.4c panel C), the defender can also perform a conical movement with its forequarters; first away from its partner, then skywards and towards its partner, to launch a counter-attack against its attacker's nape (see Pellis, 1985, for a description of the general features of this manoeuvre across species). For either of these counter-attack manoeuvres, if the recipient rears and braces itself in time, the counter-attack can be thwarted as a mutual upright configuration is achieved.

Rotation around the vertical axis involves a rat turning to face an attacker approaching from the rear (Figure 5.5). As an attacker progresses upward along the recipient's back towards its nape (a), the defender rears its body while rotating around its longitudinal axis to face the attacker (b), then rotates its pelvis so that its whole body is facing the attacker (c). Finally, from this position, the defender launches a counter-attack against the attacker's nape (d). If by the time the defender reaches the position in panel c, the attacker manages to rear to upright, the defender's counter-attack can be thwarted and the two rats gain a mutual upright position in which they grapple and box with their forepaws. Thus, both the partial rotation and the vertical rotation tactics can converge on a comparable end

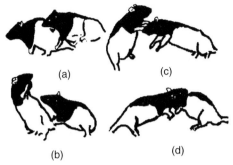

Figure 5.5 This is a form of facing defense in which the defender turns to face its attacker while pivoting on its hind legs (a, b). From this position, the defender can block access to its nape (c) and launch a counter-attack to its attacker's nape (d). Reprinted from Pellis, Pellis and McKenna (1994) with permission (Copyright © 1994, American Psychological Association)

point, that of a mutual upright. Indeed, if an attacker approaches from an oblique rear angle and the defender has already partially reared when this occurs, if it now performs a partial rotation, it is often difficult to discriminate this manoeuvre from the vertical rotation tactic. What they have in common is that they both involve standing on their hind feet. These details of the defence tactics are important because their execution can be greatly influenced by context.

As most play fights occur in corners, if the defender is pushed against the walls, its freedom of movement is greatly constrained. Evasion becomes unlikely, as either jumping forward or swerving laterally would lead the rat to crash into a wall and so not remove its nape from the proximity of its partner's snout. Similarly, while all the facing forms of defence are possible, the standing tactics would be constrained by the attacker pushing its body against its opponent's body, pressing the defender against the wall and so reducing its ability to manoeuvre to block further attacks or to position itself to launch a successful counter-attack. By being pushed against the walls, the defender is often forced onto its back. As noted earlier, the on top/on bottom position, typical of the commonly scored pin, is achieved either by the complete rotation tactic of the defender or is forced upon the defender by the continued attacking actions of its partner. In the constrained space of the corners, not only do complete rotations end in the pin configuration, but also do most other defensive tactics. By scoring all pins, irrespective of the spatial location, as being equivalent, this end point measure may be artificially inflated, implying that this is the preferred wrestling position of rats (Panksepp, 1998).

There are two solutions to this problem. First, as noted, make a judgement on the preferred tactics employed by defenders by determining the type of defence tactic initiated. This can usually be done in the first two to three video frames (Himmler, Pellis & Pellis, 2013), before protracted wrestling leads to a different end result (Himmler, Himmler, Stryjek et al., 2016). Second, as advocated in Chapter 4, only count those defences that occur in the open, away from the constraints of the walls, and only score defences to attacks from one specific direction (e.g., from the side), as these limitations would give a less biased assessment of which tactics defenders are attempting to execute (i.e., score the behaviour at decision points).

To date, all rat strains studied mostly attack and defend the nape during play fighting and use the same basic suite of defensive tactics (Himmler, Himmler, Pellis & Pellis, 2016). However, there are strain differences in the frequency of attacks and in the tactics they are most likely to use (e.g., Himmler, Stryjek, Modlińska et al., 2013; Himmler, Modlińska, Stryjek et al. 2014; Reinhart, McIntyre & Pellis, 2004; Reinhart, McIntyre, Metz & Pellis, 2006; Siviy, Baliko & Bowers, 1997; Siviy, Crawford, Akopian & Walsh, 2011; Siviy et al., 2003). Even once the confounding effects of environmental context are subtracted by the methods described above, there are still residual strain differences (Himmler, Himmler, Stryjek et al., 2016), but before it can be concluded that these differences arise from differences in how brain mechanisms regulate play, another possibility needs to be considered.

Comparing one rat strain to another is not like comparing rats to hamsters, as rats of any strain, despite variation in size and form, are still recognizably rats. For example, Sprague Dawley rats are larger than Wistar rats and compared to brown Norway rats, Sprague Dawley rats have bulkier bodies. It would seem obvious that a rat with a more lithe build would also be more agile. Indeed, even though Sprague Dawley rats and brown Norway rats are more likely to evade when defending compared to Long Evans and Wistar rats, evasion in brown Norway rats is more likely to involve leaping away, whereas Sprague Dawley rats are more likely to use a sedentary manoeuvre – that is, remain standing in place, then evade by swinging their heads and necks away from their opponents. After maturity, the sexes diverge in body size, with males becoming larger, and this difference could also affect the rats' choice of tactics when playing. However, developmental studies and comparisons across rats of different ages and strains playing together suggest that body size and even body morphology have little impact on the age, sex and strain-typical preferences in the tactics used and how those tactics are executed (e.g., Himmler, Lewis & Pellis, 2014; Pellis, Field & Whishaw, 1999; Pellis, Williams & Pellis, 2017). Nevertheless, there are small, sensory-motor differences that do have an impact on the tactics used and their successful execution (Himmler, Stryjek, Modlińska et al., 2013; Himmler, Modlińska, Stryjek et al., 2014).

Wild rats are more agile than domesticated strains and are able to leap further from a standing position. This enables them to avoid being caught in corners and, in fact, walls can work to their advantage as they can leap toward and bounce off the vertical surfaces. Not only are they able to use more acrobatic evasive manoeuvres, but they are also able to extricate themselves when caught in a wrestle by leaping away. There are also sensory differences between wild and domesticated strains. If an attacker can close the distance between itself and a defender fast enough, the defensive action will only begin once the opponent's nape is contacted. However, when not constrained by walls or an unexpected approach by a partner, a rat that is approached from its side begins its defence when its attacker's snout is about two centimetres away from its nape. This is true for domestic strains, but for wild rats, that distance is doubled – to over four centimetres. The greater distance at which defensive actions begin increases the ability of the rat to use evasive tactics as the defensive action can begin well before contact is made. Also, if defence from an approaching attacker begins at a greater distance, all tactics are more likely to be successfully executed. Multiple sensory modalities may be involved so that strain differences can remain even when one modality such as vision is blocked, as when rats are tested in complete darkness (Pellis et al., 1996).

The differences in sensory-motor capabilities suggest that simply scoring behavioural outcomes, such as pins, can greatly distort the differences across strains. This can be true even within a strain, between the sexes. When we first described the vertical rotation tactic in Long Evans rats, we found that females were more successful in executing this tactic, with success judged by whether the defender avoided having its nape contacted and managed to counter-attack its attacker's nape successfully (Figure 5.5). The reason for this difference is that females initiate the defensive tactic sooner than males. By initiating this defensive tactic a little later, a male allows its attacker to move higher up, onto its back, so that, when the defender begins to turn, the attacker can then push him to the ground (Pellis, Pellis & McKenna, 1994). Again, because of these sensory-motor influences on which tactics are used and how effective they are when they are used, to compare across strains, ages and sexes, a standardised context needs to be evaluated in which the effects of these influences are minimised (Himmler, Pellis & Pellis, 2013).

The Limits of Some Currently Used Measures of Play

Having evaluated the play fighting of rats by integrating the principles outlined in Chapters 2–4, the problems created by many of the standard ways in which play is scored (Blake & McCoy, 2015) can now be re-evaluated. The basic problem with many scoring schemes – be they an index derived from counting several behavioural markers or consist of a single, highly reliable measure, such as the pin – is that they combine the influences of attack and defence. Consequently, as already described in detail for pinning, when there is a change in the frequency of this key measure, the researcher cannot determine whether that change results from a change in the frequency of attack, the likelihood of defence, or from a change in which defence tactics are most likely to be used. In part, the addition of scoring the frequency of nape contacts along with pins (Panksepp, 1998) has helped reduce some of this ambiguity. If an increase in pinning is associated with an increase in nape contacts of similar magnitude, then it can be assumed that the change in the frequency of pinning arises from an increase in the frequency of nape contacts. As already noted, however, nape contact is also a mixed measure, incorporating the attack and defence behaviours of both partners, so it can only be a partial solution to the problem. However, even if the more inclusive measure of nape attack were used, what if the magnitude of the change in the frequency of pinning is not similar to that of nape attacks, or even worse, what if the direction of the change in the two measures are opposite? These ambiguous outcomes require knowledge about possible changes in the likelihood that the rats defend themselves and changes in the use of particular defensive tactics. For example, the rats could be less likely to use tactics that lead to a pin. As any particular experimental manipulation that changes the frequency of pinning could do so by differentially altering the frequency of attack, the likelihood of defence and the type of defence used, it is not possible, a priori, to predict if a singular measure will be sufficient to identify the causal processes leading to a change in that measure (Himmler, Pellis & Pellis, 2013).

The details in how particular actions are measured can also have unpredictable consequences. Nape contact is defined as one rat placing its forepaws on its partner's head and shoulders while contacting its nape with its snout (Panksepp, 1998). But as noted, with regard to the sensory-motor influences on the use of playful tactics, when rats are unconstrained by their surroundings, they can initiate a defence before their nape is contacted. In some cases, such as for wild rats – which begin to defend at an even larger distance than most domesticated strains and also have considerably more acrobatic ability – actual nape contact is rare. This problem can be overcome by tracking the animals' movements to determine whether the defender begins its defence as its partner is approaching its nape.

When an attacker approaches from its opponent's rear at an oblique angle with its snout pointed at their nape, this elicits a defensive response by the defender. However, if an attacker approaches from its opponent's front at an oblique angle with its snout pointed at their rump, it does not. Thus, while being approached by one's partner in close proximity may be considered a cue of imminent social contact by human observers, the recipient of that approach informs us that it regards the former as a playful attack and the latter as not. Recall the importance of using clues from the animals themselves to determine what constitutes their *Umwelt*. The oblique frontal approach to another's rump may signal imminent social contact important for social investigation, derived from sniffing the anogenital area. However, social investigation contact is not defended (Panksepp, 1981), but nape contact is (Pellis & Pellis, 1987; Siviy & Panksepp, 1987). The oblique rear approach with the snout targeting the nape does seem to be interpreted by a recipient as signalling imminent nape contact. That these are real discriminations made by the rats themselves is shown in their vocalisations.

During play fighting and other social interactions, rats emit ultrasonic vocalisations (Burgdorf et al., 2008; Himmler, Kisko, Euston et al., 2014). Comparing the variety of such calls (Wright, Gourdon, & Clarke, 2010) with particular behavioural actions show that, while imminent contact involving social investigation is not associated with calling, particular calls are emitted with imminent contact of the nape (Burke, Kisko, Euston, & Pellis, 2018). Approaching the nape is associated with calls that are also emitted when the nape is contacted, but when the anogenital area is approached – usually preceding social sniffing – such calls are not emitted (Pellis, Burke, Kisko, & Euston, 2018). By only scoring nape contacts, the play fighting by rats can be underestimated, as the rats themselves have a broader view of what constitutes a nape attack. Therefore, scoring nape attacks, the initial defensive actions attempted and how the combined behaviour of the pair give rise to composite measures such as pins provides a framework that more closely reflects the organisation of play fighting (Himmler, Pellis, & Pellis, 2013; Himmler, Himmler, Stryjek et al., 2016).

Principled Approaches Can Lead to Novel Insights and Measures

With appropriate modifications due to the behavioural system being simulated, environmental context, and bodily factors, play fighting can be understood as being an interaction of attack and defence of the nape. If so, then all actions performed by the rats should be understandable when viewed through the lens of compensation; an action by one animal should be present because it blocks or overcomes the actions of the other animal. This competition for the nape made sense of much of what we saw in the play fighting of rats (Pellis & Pellis, 1987) and is the basis for most of the behavioural markers that we have recommended for scoring play quantitatively (Himmler, Pellis & Pellis, 2013). But applying this framework for scoring play fighting in rats led us to notice some anomalies in their behaviour, which have required us to modify our conception of how play fighting is organised. In turn, these new insights have led to the development of new behavioural markers that capture these anomalies of organisation quantitatively.

Several of these novel insights and measures have been presented elsewhere (e.g., Himmler, Himmler, Pellis & Pellis, 2016; Himmler, Himmler,

Stryjek et al., 2016; Pellis, Williams & Pellis, 2017), so for the purposes of illustration we will focus on one. As noted in earlier chapters, during serious fighting, animals compete to bite or strike their opponent without themselves receiving a bite or a strike (Blanchard, Blanchard, Takahashi & Kelley, 1977; Geist, 1978). That is, whether one animal gains the upper hand and mercilessly pounds another animal or one animal prevents every attempt by another animal to gain an advantage and so produce a stale-mate, both would equally qualify as serious fights (Geist, 1978; Pellis, 1997; Pellis et al., 2013). In contrast, even though play fighting involves compet-ing for an advantage (Aldis, 1975), if either of these two outcomes occurs in the context of a playful encounter, it would likely lead to one or both partners abandoning the interaction. That is, for play fighting to remain playful, animals have to have some degree of reciprocity, so that both partners have at least some opportunity to gain the advantage (Palagi, Cordoni, Demuru & Bekoff, 2016). For example, when one partner uses excessive force and continually overpowers its playmates, those partners cease engaging that individual in play (Suomi, 2005; Wilmer, 1991). Some level of turn-taking in play is essential, although the degree of such reciprocity needed and how animals engineer opportunities for their partner to have a chance to gain the upperhand varies with age, sex, dominance relationships, and species (Pellis & Pellis, 2017). Thus, while both forms of fighting involve competition, the competition present in play fighting is moderated by cooperation.

In serious fighting, the correlation between attack and defence can explain the majority of actions observable to a human witness. However, because of the presence of cooperation in play fighting, this relationship between attack and defence is looser and can result in actions that are not explicable in terms of attack and defence. For example, we have already seen that, in rats, a common configuration for the partners is the pin (see panel h in Figure 5.1). While this is a defensive position for the rat on the bottom, as its nape is held away from its partner's snout, the rat on top has the advantage, as it can use its forepaws to hold and block its partner's ability to counter-attack. By being in the position on top, the rat can maintain its postural stability by having its hind paws planted firmly on the ground, which allows it to use its forepaws and upper body to counteract its partner's movements (Figure 5.6a). Rats also do this during serious fighting (Blanchard, Blanchard, Takahashi & Kelley, 1977). However, during play

(a)

(b)

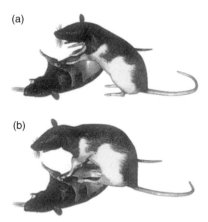

Figure 5.6 When engaged in play fighting, rats often adopt a posture in which one animal is standing over the other rat, which is lying on its back. However, the posture of the rat on top can take one of two forms: while standing on the ground with its hind paws, it can hold its partner down with its forepaws, (a) or it can stand on its partner with all four of its paws (b). Reprinted from Foroud and Pellis (2003) with permission (Copyright © 2003 John Wiley & Sons, Inc.)

fighting, rats sometimes adopt a posture that makes no sense at all with regard to gaining and maintaining the advantage – the rat on top may stand on its supine partner with all four of its feet (Figure 5.6b).

In this position, the rat on top has difficulty maintaining its postural stability as the rat on the bottom squirms. Indeed, chances of a successful counter-attack aimed at the nape from the rat on the bottom when the rat on top is standing with its hind feet anchored on the ground is around thirty percent, whereas, when in the unanchored position, the success rate jumps to over seventy percent (Pellis, Pellis & Foroud, 2005). That is, during play fighting, rats engage in movements that compromise or handicap their ability to maintain the advantageous position, movements that give their partner an opportunity to reverse the situation. While undermining the animals' own advantage, these manoeuvres, inserted into the play, increase the opportunity for the play fights to become reciprocal. Scoring such movements can be daunting, but as their presence affects the likelihood that during play partners reverse roles, the role reversals (in which the original attacker becomes the defender) themselves, which are easier to observe, can be scored as a proxy measure (Himmler, Himmler, Pellis & Pellis, 2016). Changes in the occurrence of role reversals can inform observers about the relationship between the partners and the consequences of the experiences derived from playing (e.g., Pellis, Williams & Pellis, 2017; Pellis et al., 2019). The methodological lesson to be derived from this example is that, by explicitly using testable principles in describing behaviour and abstracting behavioural markers to score that reflect the underlying organisation of the behaviour (Chapters 2–4), anomalies that are inconsistent with those principles can be identified. In turn, such anomalies can alert us to unexpected aspects of organisation that then lead to characterising new behavioural markers that can be used to score that newly discerned novelty in organisation.

New ways of testing proposed behavioural markers for how well they reflect the underlying behavioural organised would be helpful, as would the identifying of new principles that may shape how behaviour is organised. Some possibilities that could greatly refine future behavioural description and the abstracting of new and more accurate behavioural markers are considered in Chapter 6.

6   What of the Future?

(T)he great tragedy of science is the slaying of a beautiful hypothesis by an ugly fact

Thomas H. Huxley (1901, p. 508)

A pack of wolves hunts in a coordinated fashion, suggesting complex, cognitive evaluations of what both the prey and the pack mates are doing (MacNulty, Mech & Smith, 2007). But how complex do the evaluations by the wolves need to be to produce such coordination? Using agent-based modelling – a computer simulation approach that creates virtual subjects that follow programmed rules (Railsback & Grimm, 2011) – a study concluded that it takes little evaluation by the subjects to produce the coordinated behaviour among virtual wolves (Muro, Escobedo, Spector & Coppinger, 2011). They showed that each wolf in the pack need follow only two rules: (1) track and close the distance with the prey and (2) maintain a constant spatial relationship (relative position) with the wolf closest to them. That is, by following two perceptual constants (see Chapter 2), the wolves are able to produce the seemingly complex coordination between pack members during the hunt.

This example highlights two important issues related to ensuring that the behavioural markers used for quantitative scoring closely reflect the organisation of behaviour. First, technologies like agent-based modelling allow researchers to test whether rules of organisation discerned from experimental and observational studies are capable of producing the observed behaviour. Prior to such methods, additional experiments and observations of the animals themselves were the only avenues available. Typically, we have followed this approach; if the animals are following rule X, then to compensate for a particular disturbance, they should produce behaviour Y (Pellis et al., 2013). Similarly, the strength of the rule can be tested with species, age, and sex comparisons (Pellis & Pellis, 1998; Pellis,

Field, Smith & Pellis, 1997). Creating virtual animals provides another approach for such testing. If the rules used to generate behaviour in virtual animals produce data that are consistent with data derived from experimental, comparative, and observational approaches, it adds to the converging evidence for the likely veracity of the proposed organisational principles. Conversely, failure of those rules to recreate the behaviour of real animals casts doubt on whether those rules are necessary and/or sufficient, prodding the use of more conventional analyses to probe the phenomenon further. The refined rules can then be re-tested with virtual animals in an iterative process.

Second, the model developed for wolves hunting in a pack provides a way of explaining the variability in the behaviour that is scored. According to this model, the observable behavioural output of the wolves is accounted for by the compensatory actions taken by each wolf as it regulates its movements relative to the prey and its nearest wolf neighbour. In tracking the movements of a particular wolf, changes in its trajectory of movement and its speed should be explainable by the dynamics of its relationship with its prey and fellow wolves. That is, variation in behaviour is explainable by the factors that contribute to the dynamics of the interaction (see Chapters 2, 3, and 4). From this perspective, an imperfect explanation for the variation in the behaviour would reflect our ignorance of the dynamics involved. For example, in his observation of a dog hunting a quail, Vladimir Krushinsky noted that when the quail ran into a bush that was too thick for the dog to penetrate, the dog circled around the bush, apparently to wait for the quail to emerge at the other side (Dugatkin & Trut, 2017). In this case, simply following perceptual rules tracking the prey is insufficient to account for the dog's behaviour. Some additional rules may be needed to account for this apparent anticipatory behaviour (see Pellis & Bell, 2020; Poletaeva & Zorina, 2015).

There is, however, an alternative view: sometimes randomness – and so unexplained variation – is an asset. Lynx hunt hares, chasing them across open spaces. As part of their defensive manoeuvres, the hares make unpredictable changes to their path of running, a sharp turn to the right here, a sharp turn to the left there. These protean, or erratic, movements increase the difficulty for the lynx to reach or intercept a fleeing hare (Driver & Humphries, 1988). In the movie, *The Hunt for Red October* (1990), the main protagonist, Jack Ryan, claims that he knows the captain of the enemy submarine so well that he can predict the direction of the next 'crazy Ivan' (a protean change in the direction of movement). He does this to convince the captain of the American pursuit submarine to trust him by claiming that the enemy captain always veers to the left at the bottom of the hour. Ryan is correct, but it was actually a guess; he didn't really know. If he did, then so would the American captain, as, after a few such crazy Ivans, it would become clear that the Russian captain is following this simple rule, and not behaving randomly. The only way to make protean manoeuvres foolproof is to make them as random as possible. The problem is, though, that of how the brain can produce random movements.

Playing rock, paper, scissors requires that you guess your opponent's choice, so to succeed you need to be as random as possible. This is a surprisingly difficult ask. A simpler example will illustrate the problem. Consider being at a computer terminal and being tasked with typing either '0' or '1' to produce a string of numbers but with one instruction 'be as random as you can'. As the length of the string increases, a non-random pattern will begin to emerge. Great players of rock, paper, scissors are adept at detecting the patterns in others and in producing choices that are more difficult to predict; that is, at making more random choices (Hood, 2010). Indeed, even mechanical random number generators are not completely random, since, with enough trials, non-random patterns become detectable (Park, 2000). It is unclear how randomness is produced and how random it may actually be (Sandhu, Gulrez & Mansell, 2020), but in real-life situations, animals do not have the luxury of statistically evaluating dozens, hundreds, or thousands of trials to detect a non-random pattern. Life-and-death decisions have to be based on the immediate context (Ellis, 1982); expending much cognitive capacity to discern a pattern or produce a random action (Hood, 2010) is not a winning option.

The hare's brain does not need to produce truly random movements, just movements that are random enough to make it difficult for the lynx to predict a turn to the right or left. If the hare's protean movements are too easy to predict over the course of a couple or so trials, natural selection would eliminate that hare, leaving to reproduce hares ones that are capable of making more difficult-to-anticipate manoeuvres. From the point of view of how to characterise the movements of the hare, which is the concern of this book, the above considerations suggest that knowing the movements of the lynx would not be sufficient to account for all the hare's behaviour. The hare interjects relatively random movements that are independent of the movements made by the lynx. The use of such random processes, however imperfect, may be important for a variety of social and non-social behaviours (Miller, 1997; Wainwright, Mehta & Higham, 2008), and such randomness needs to be integrated with the processes presented in Chapters 2, 3, and 4.

Testing Descriptive Hypotheses with Virtual and Artificial Animals

In their first week or so of life, rat pups are pink and hairless and are not, as yet, able to self-regulate their body temperature. Thus, when their mother leaves them, their body temperature drops. To compensate for this cooling, the pups huddle together (Figure 6.1). Experimentally, this huddling can be studied by placing six to eight pups in an open field. Shortly after this, they will begin to move and eventually will cluster into a huddle. How do these isolated pups move so as to contact their siblings until they all are grouped together? This is a particularly interesting question, because, for their first ten to fourteen days, rat pups' eyes and ears are closed and their vibrissae are either not present or are very short. That is, their long-to-mid-distance sensory systems are not available to guide their behaviour. By doing

Figure 6.1 Groups of huddling rat pups are shown at two ages: 5 days and 20 days. Unlike the younger pups (bottom group), the older pups have their eyes open and their bodies are furred. From Alberts, J. R. (2005). Infancy. In I. Q. Whishaw & B. Kolb (Eds), *The behavior of the laboratory rat. A handbook with tests* (pp. 266–277). New York: Oxford University Press. (Copyright © 2005, Oxford University Press, reproduced with permission of the Licensor through PLSclear)

detailed experimental manipulations in which the humidity and air temperature were controlled along with the shape, texture, and temperature of the surfaces of the test enclosures, Jeff Alberts concluded that huddling behaviour involves the rat pups following some simple rules (see Alberts, 2007, for an overview). The pups move about, but once they make contact with a vertical surface, they maintain that contact. They also show a preference for warmer surfaces. Since the only vertical surfaces available in the test enclosure are the walls and other pups, and neighbouring pups are warmer than the walls, once they contact another pup, they remain in contact. The process continues until all the pups in the test enclosure are huddled together. Up until the seventh day, rat pups are insensitive to the activity of adjacent pups; they simply move to maintain contact with the 'warm vertical' surface. However, after this age, the level of activity by the pups is influenced by the activity of their littermates, with them becoming more or less active depending on the activity of their neighbours. Experimentally, following the two rules about vertical surfaces and warm surfaces seemed sufficient to account for the behaviour of pups in their first week of life, but then to account for the movements of pups in their second week of life, a rule involving modulating one's own activity in relation to the activity of one's neighbours had to be added.

Joined by Jeff Schank, who has expertise with agent-based modelling, the two Jeffs created a model that simulated the behaviour of huddling pups. In the program, pups were represented as short sticks on a computer screen, and eight of these sticks were placed in a virtual enclosure circumscribed by walls. The virtual pups were programmed to move according to the three rules. When only the first two rules were activated, the cluster of sticks mimicked the behaviour of pups at seven days old or younger, but when all three rules were employed in the model, the sticks mimicked the behaviour of the older pups (Schank & Alberts, 1997).

There are two limitations with these virtual animals: they are two-dimensional sticks, and so do not have the three-dimensional structure of real pups, nor do they have the flexible bodies of real rats. Thus, even though the computer-generated sticks are limited by the boundaries of their environment and can only move according to the programmed rules, the effects of body morphology cannot be assessed (see Chapter 4). Jeff Schank and his group tackled this problem in two ways. In one approach, the virtual animals were programmed to have differing body morphologies, with one, two, or three bendable components. Simulating the huddling behaviour of seven- and ten-day old pups following their respective suite of rules showed that the behaviour of the seven-day old pups was best explained when there were three segments to their bodies. In contrast, varying body morphology had little impact on the behaviour of the simulated ten-day old pups. This simulation identified the behaviour of the younger pups as being more greatly constrained by body morphology than that of the older pups (May, Schank & Joshi, 2011). In the other approach, Jeff and his team constructed small, metal robots that moved on wheels, thus having real three-dimensional structure (May et al., 2006; Schank, May, Tran, & Joshi, 2004). Scaling the robopups down to the same size as real pups was not possible, but scaling the test arena up to reflect the larger size of the artificial pups meant that the aggregation behaviour of real and artificial pups (Figure 6.2a) could be compared. Again, when programmed with the rules followed by the real animals, the behavior of these 'robopups' resembled the behaviour of real pups to a remarkable degree, achieving very similar aggregation patterns (Figure 6.2b). Thus, using computerised simulations and robotics, the researchers were able to demonstrate that the rules derived from the analysis of live pups were necessary – but not sufficient – to account for all the pups' movements. Physical surfaces, body morphology and random processes were also involved (May & Schank, 2009).

The example given demonstrates the new insights to be gained from an iterative process of exchange between observing, experimenting, and modelling (Alberts, 2012). Virtual animals can be useful not only for testing hypotheses about the perceptual rules guiding behaviour (Chapter 2); they can also provide a realistic means by which to assess the impact of environmental context and body morphology on constraining the expression of those rules (Chapter 4). Robots that have real three-dimensional structure that can physically interact with their environment may provide particularly compelling opportunities to test and refine our conceptions of how behaviour is organised (Webb, 2001; Webb & Consi, 2001).

Such imagined animals can also provide a way of testing whether built-in biases for motor action – programmed in the brain – are needed. For example, dodging away from an opponent laterally is a manoeuvre seen in protecting a food item from a conspecific (see Figure 4.5), but it is also present in sexual, agonistic, and predatory encounters (Pellis & Bell, 2020). One way to interpret the organisation of lateral dodging is that, like the

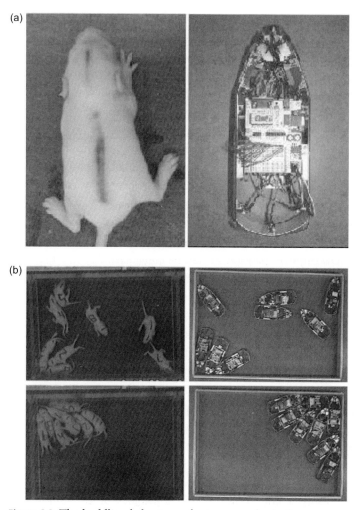

Figure 6.2 The huddling behaviour of rat pups and robotic pups (robopups) is compared (a). The real and artificial pups were released into their respective arenas (b top panels) and then allowed to move about. The real pups and robopups achieved many comparable patterns of aggregation (b bottom panels). Adapted from May et al. (2006) with permission (Copyright © 2006 John Wiley & Sons, Inc.)

various hand and body movements of monkeys (see Graziano, 2009, 2016 and Figure 3.2), it is a motor primitive (Flash & Hochner, 2005) – a built-in motor action that can be integrated into many different behavioural contexts. With regard to its use in protecting food, contextual factors like the quality of the food, the speed of approach, and the imminence of contact can be integrated into an algorithm that calculates the magnitude of the movement (Whishaw & Gorny, 1994b). But does the brain need to have the dodge movement embedded in its structure? By taking the analytical perspective characterised in Chapter 2, that of identifying what perceptions the animals are keeping constant (Bell & Pellis, 2011), and then applying agent-based modelling (Bell, Bell, Schank & Pellis, 2015), a possible answer to this question can be derived (Bell, 2014; Pellis, Pellis & Iwaniuk, 2014).

Once the robber reaches a threshold distance between its snout and that of the rat holding the food item (on average, about two centimetres), the food holder, or defender, dodges laterally (see Figure 4.5) and continues to move laterally until it reaches and maintains a specific distance between its own snout and that of the robber (on average, about ten centimetres). Consequently, if the robber stops its movement once the dodge is initiated, the total magnitude of the angle of the dodge is less than if the robber pursues the retreating defender. This leads to a positive correlation between the movement of the robber and the movement of the defender (Figure 6.3a). However, the amount of movement made by the defender is in the service of gaining and maintaining a constant distance between the snouts of the two animals (see Chapter 2). Thus, irrespective of the amount of movement performed by the robber, the inter-snout difference remains unchanged (Figure 6.3b). The quality of the food can change the magnitude of the inter-snout distance that is gained and maintained, but it does not change the perceptual rule being followed (Bell & Pellis, 2011).

An agent-based model was developed for the robber–dodger paradigm with the virtual rats being governed by two rules – first, when the opponent's 'snout' comes within a certain distance of the food owner's 'snout', move away, and second, continue moving until a particular inter-snout distance is gained and maintained (Bell, Bell, Schank & Pellis, 2015). Two things emerged from the behaviour of these virtual animals. First, for many of the robber's approaches, the defender dodged laterally (Figure 6.4). Second, these dodges mirrored the pattern shown by the real animals; there was a positive correlation between the amount of movement made by the robber and the defender, but there was no correlation between the inter-snout distance and the amount of movement made by either the robber or the defender. For our purposes, the computer model revealed that the virtual animal did not need to be instructed to dodge laterally; the direction of approach and the creation of an appropriate inter-snout

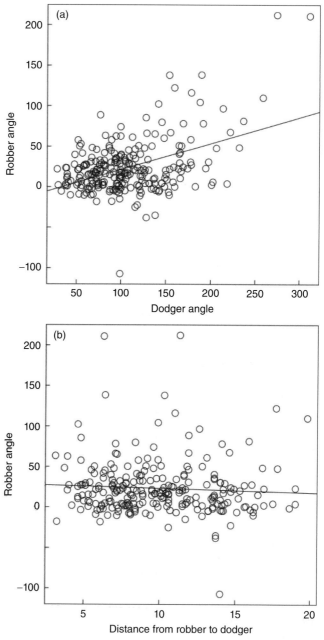

Figure 6.3 Two correlations depict the relationship between the rats defending their food by dodging away from robbers. In (a), the amount

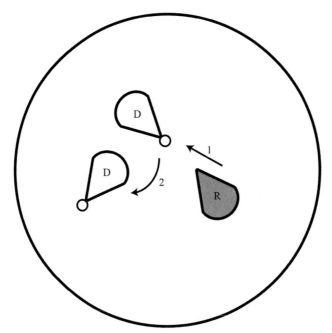

Figure 6.4 This drawing is from a simulation of dodging, with the virtual organisms having a front end (pointed part) and a back end. The defender – triangle D – is shown in position 1 holding the virtual food item, as the robber – triangle R – approaches from the front. As shown in position 2, once the robber closes in to the threshold distance, the defender pivots away laterally. That is, the virtual defender performs a dodging manoeuvre in order to protect the food item. Adapted from Bell (2013) with permission from the author

Figure 6.3 (*cont.*)
of movement by the defender dodging is positively correlated with the amount of movement by the robber (r = 0.449, p < 0.001). In contrast, as shown in (b) the amount of movement by the robber is not correlated with the inter-snout distance between the two rats (r = –0.059, p = 0.360). Reprinted from Bell and Pellis (2011) with permission (Copyright © 2011 Elsevier Ltd.)

distance was sufficient to produce a lateral dodge. That is, the model suggests that the rat's brain does not need to have a built-in motor rule to produce this particular movement.

With further developments in computing and robotics, the future use of virtual or artificial animals to test hypothesised rules that govern the behaviour of real animals has great potential. In particular, incorporating contextual factors and differences in body morphology into such models may lead to being able to identify more realistically not only the common rules governing the behaviour of divergent animals, but also to characterise the reasons for the dissimilarities (e.g., Bell et al., 2012). Also, while the example for lateral dodging indicates that no motor instruction is needed, the tactic is relatively simple, requiring the animal to move in only two dimensions. More sophisticated models or robots (e.g., May, Schank & Joshi, 2011; Schank, May, Tran, & Joshi, 2004) may also be extended to more complex tactics that include movements in three dimensions involving multiple and variable combinations of movements by different body parts (e.g., Pellis, 1985). The more complex the movement sequence needed to complete a task, the more likely it is that there are built-in motor biases, as seen earlier for the various types of righting (Pellis, 1996). More realistic virtual or robotic animals could be used to test the need for such inferred biases.

Where Does a Stochastic Principle Fit into the Descriptive Pantheon?

Rats with damage to the lateral hypothalamic area of the brain remain akinetic (i.e., immobile), but their capacity first to stand and then to move about gradually recovers. When the movements in various dimensions – vertical, lateral, and forward – are present, the rat can still become immobile or 'trapped' in a tight corner (Golani, Wolgin & Teitelbaum, 1979). By being limited by a pre-specified, geometric sequence of movement, the rat may begin to move out of the corner, but then move back in (Figure 6.5). Thus, even when having a full complement of motor actions, if those motor actions are relatively insensitive to the environmental cues, they may not be sufficient to navigate a world with obstructions. Having the guidance of sensory cues is essential.

Even so, following sensory rules too rigidly can also create problems. Rodents like mice and rats are thigmotactic (Barnett, 1975), so when they

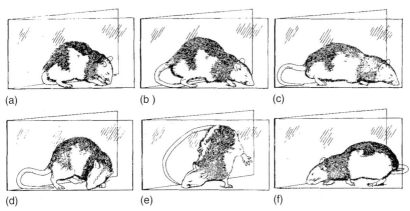

(a)   (b)   (c)

(d)   (e)   (f)

Figure 6.5 In drawings made from frames of movie film of a rat with damage to the lateral hypothalamus, it is shown being trapped in a corner after it has seemingly recovered from being akinetic. After moving to the maximum extent along one dimension, such as in the forward plane (a–c), the rat then changes direction (d–f), trapping itself in the corner. Adapted from Golani, Wolgin, & Teitelbaum (1979) with permission (Copyright © 1979 Elsevier Ltd.)

begin to explore a new area, they maintain tactile contact with a vertical surface (e.g., the surrounding wall) with the vibrissae on one side of their face, resulting in the shape of the vertical surface determining the path of locomotion (Golani, 2012). But what happens if a rat encounters a corner in a square or rectangular enclosure, in which the vibrissae on both sides of its face have equal contact? How does the rat break the symmetrical contact to continue its path of exploration? An example from an experiment on brain-damaged rats provides a clue.

Brainstem circuits are involved in generating locomotion (Grillner, 1975). Damage to one particular region, the *nucleus reticularis tegmenti pontis* (NRTP), causes rats to accelerate forward rapidly (Cheng, Schallert, DeRyck & Teitelbaum, 1981). That is, when the rats begin to walk forward, they quickly accelerate to a gallop, only stopping when they strike the wall with their head (Figure 6.6). Once in contact with the wall, the rat will remain immobile, and will only resume its forward locomotion if the experimenter moves the rat so that the wall no longer provides symmetrical contact on its face. Indeed, the uncontrolled forward locomotion can be stopped completely by wrapping an elastic band around the rat's snout

(a)

(b)

(c)

(d)

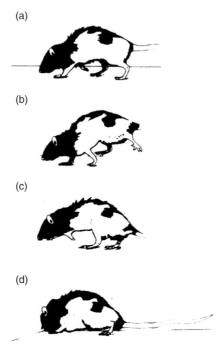

Figure 6.6 Drawings traced from frames of movie film show the accelerating gait of a rat with NRTP damage. The rat begins by walking (a), then running (b) and then galloping (c). However, once it crashes headlong into a wall (d), all locomotion is shut down. Adapted from Cheng, Schallert, DeRyck, & Teitelbaum (1981) with permission from the contributing author, Philip Teitelbaum

(Teitelbaum, Schallert & Whishaw, 1983). The NRTP damage is somehow releasing the rat from the control of higher-level brain mechanisms that normally restrain uncontrolled forward locomotion (Sinnamon, 1993) and also from the ability to overcome the incapacitating effect of symmetrical stimulation on its face.

If intact rats are limited to following two rules – keep contact on one side of one's face and walk forward when contact is asymmetrical – then they would end up like the NRTP rat and stop dead in their tracks when the contact on their snout is symmetrical. Again, virtual animals can shed light on what may be missing in the NRTP rat. Bill Powers developed computer simulations in which the virtual organisms followed such simple sensory rules and he discovered that, like the NRTP rats, they became stuck in corners when the contact was symmetrical. The way that he produced virtual animals that could escape from such a trap was to introduce another rule: when contact is symmetrical, make random movements. The random movements broke the symmetrical contact and allowed the organism to begin its forward locomotion again (Powers, 2009). In this case, the random element in movement is not noise that slips through our ability to explain, but rather, forms a crucial component of adaptively navigating the world. This phenomenon is well known in the movements of bacteria; they move away from toxic chemicals and toward chemical cues indicating food. But when such sensory inputs are not present to guide its movement, after some period of swimming in a particular direction, the bacterium stops, goes into a tumble and then sets off in a new direction (Adler, 1975). Physical factors, such as light levels and oxygen concentrations (Armitage, 1997), and even the presence and activity of other bacteria (Lyon, 2007), can all influence the frequency and duration of the tumbling, but the change in the direction of movement is essentially random.

Coupled with an adaptive role for random movements, in some situations, another source of inexplicable movements arises from the imperfection of biological systems. The imprecision at all levels of organisation leads to unpredictable noise in the system. That is, even when the context is invariant, behavioural output often remains variable (Wainwright, Mehta & Higham, 2008). Again, artificial animals provide some insight. Robots in a factory line, with each performing a specific task, are programmed to execute highly precise movements. Applying this conception of a robot to machines that are capable of being autonomous, in being able to traverse an unknown landscape, turned out to be a failing strategy. Rather, autonomous robots had to be equipped with two properties (Pfeifer & Bongard, 2007). First, robots had to follow perceptual constants, allowing for variable behaviour to maintain that constancy. Second, and most directly related to the present musings, there had to be some slack in how the component parts articulate with one another. If the connections were too rigid, postural stability is reduced, impairing the artificial organism's ability to adapt to variable terrain. The implication for real animals is that some level of randomness in behavioural output at the level of how the brain encodes movement and in how the body executes movements is likely a central feature of adaptable behavioural organisation. Indeed, such random production of movements may help in learning new motor skills requiring novel combinations of actions (Herzfeld & Shadmehr, 2014).

Of course, experimentally demonstrating that a given variant in behaviour does not arise from some pre-wired option or from dealing with some environmental cue not obvious to the researcher is no easy feat. Critics can always claim that there is an adaptive reason for the variant that has not as yet been identified. Still, there are some empirical studies that seem to have made such extraneous factors highly unlikely, leaving the strong possibility that the variability present is beyond what is needed by the exigencies of the context (e.g., Berridge & Fentress, 1986; Eberhard, 1990). As indicated by the building of autonomous robots, in some situations, imprecision or noise in the execution of behaviour may actually be beneficial in allowing animals to deal with a changing and unpredictable world. If this is true, no matter how well we understand the perceptions that are relevant to an animal (Chapter 2), how the behaviour engaged is organised (Chapter 3) and how body morphology and context may constrain and modify behavioural output (Chapter 4), there will be residual variability present. At least some of that variability may not simply be due to our ignorance of the above factors and how they interact, but instead, be a product of intrinsically stochastic processes. Identifying behavioural markers that reflect such a stochastic organisational principle would be a major step forward in the future of behavioural analysis. New computational methods that can deal with large quantities of data, while they may have their limitations (see Chapter 1), may be critical in such an endeavour (e.g., Anderson & Perona, 2014; Brown & de Bivort, 2018). So would characterising how all the principles discussed in this book, including the stochastic one considered in this chapter, are integrated together by meta-principles. This is the gauntlet we throw down for the next generation.

Epilogue

If you have made it to here, we hope that you have enjoyed the journey. The test of the value of the journey is whether, as you gained some new insights, you looked at your pet cat or dog or at animals at a zoo in a different way. Most importantly, if you are contemplating a scientific study, hopefully you have acquired some new insights into the process by which to decide what may be the most profitable aspects of the behaviour to measure. What we have offered is a glimpse into the factors that contribute to how behaviour is organised and have, hopefully, shown how abstracting behavioural markers can be rendered into a more object-ive process. Even if derived from a formal application of the principles that underlie the organisation of behaviour, once multiple researchers apply them, some behavioural markers will weather the test of time but others will not. In the latter case, the proposed markers will be found to be poor reflections of the organisation they are meant to represent. That is what science is all about; some hypotheses stand up to scrutiny and some do not.

For us, after over thirty years of repeated observation, the hypothesis that most of the behaviour present in the play fighting of rats can be understood as a competition over access to nuzzling another's nape has been substantiated (Pellis & Pellis, 1987; Pellis et al., 2019). But even this is only partially true. We have good evidence that, at least in some strains of rats, around five percent of the playful interactions involve competition for nipping the rump (Modlińska, personal communication, August 2018). That is, a minority of playful attacks is directed at the body area targeted during serious fighting (Blanchard, Blanchard, Takahashi & Kelley, 1977; Pellis & Pellis, 1987). Also, as shown in Chapter 5, some actions performed during play fighting, especially those related to self-handicapping, are not consistent with attack and defence of either the nape or the rump. That is, even with such a well-studied behaviour in an intensively observed species, the description of what the animals do is an evolving one, and, along with the revised descriptions, different behavioural markers are needed to reflect that new understanding.

This is the main lesson we wish readers to gain from this book. However principled the selection of behavioural markers, as one's knowledge of the behaviour in question expands, the applicability of those markers will need refinement in how they are measured. As behavioural description is an evolving process, it is worth re-stating the core steps in how to describe behaviour and then how to extract suitable behavioural markers for quantitative analyses.

   I. Systematically apply the principles of behavioural organisation (Chapters 2–4) to the unfolding description of the behaviour that you are studying.
  II. Select behavioural markers that reflect the organisation revealed by the descriptive analyses.
 III. Given that the selected behavioural markers can be considered as hypotheses (i.e., they reflect the causal processes discerned from the description), they need to be tested for their explanatory power. If they pass the test, then they can be used with confidence to quantify the behaviour being studied.
 IV. If the test reveals that the selected markers are a poor reflection of the underlying behavioural organisation, then reject them. Similarly, if the behavioural markers are reasonable, but only incompletely reflect the underlying organisation, then they need to be augmented or refined.
  V. Inadequate or only partially satisfactory behavioural markers require that the behavioural description be renewed as something is clearly missing. Two reasons with which we are familiar could be involved: (A) The known principles of organisation may have been inadequately applied or their interaction not sufficiently taken into account; (B) some new organising principle is at play and it needs to be identified and characterised so that it can be added to the workings of the other known principles.
 VI. Once a deeper understanding of the organisation of the behaviour is gained, select new behavioural markers and try applying them again.

This iterative process continues until the researchers have developed behavioural markers that they are confident are suitable for quantitatively testing the aspects of the behavioural organisation that they are studying (see Appendix B).

Basically, we are advocating for the development of quantifiable behavioural markers that arise from a process, one that conforms to the standards of the scientific method. Selecting a behavioural marker is positing a hypothesis that the marker arises from the causal processes that underlie the behaviour being studied. As such, that hypothesis needs to be tested. By starting with organisational principles that have had historically demonstrated value in their application (e.g., Ewert, 2005; Glimcher, 2003), we hope that this book gives readers a way with which to begin the iterative process we advocate. As noted at the end of Chapter 6, what we present in this book is only a start to a process that will add new organisational principles and novel analytical tools to gain a deeper understanding of how behaviour is produced. Thus, this book is the beginning of the journey, not the end.

Appendix A: Eshkol-Wachman Movement Notation and Descriptive Analysis

A fundamental problem in describing animal behaviour is seeing behind the veil – that is, seeing past the flurry of movements that catch our attention but that may mask what is critical from the animal's perspective. Thus, it may not be surprising that some of the best descriptions have come from researchers who have studied many species – variations that allow identifying common themes (e.g., Leyhausen, 1979), have spent time trying to draw or paint animals engaged in a variety of behaviours (e.g., Walther, 1984), or pioneered filming behaviour at a time when costs were high requiring attaining the intuition necessary to identify the key moment to press the record button (Tinbergen, 1951). We have dabbled in all the above, but having limited artistic ability, we found another method that has been useful in sharpening our observational skills, using notation systems that provide a disciplined and formal structure within which to observe and record movements.

We are familiar with two forms of hand-written choreographic notations that can be used to reveal the organisation of behaviour. The first, Laban Movement Analysis, is a universal language for human movement that records qualitative features of movement into structured and quantifiable categories (Hutchinson, 1977). It has been successfully used to notate the behaviour of both human and non-human animals for scientific analyses (e.g., Fagen, Conitz & Kunibe, 2000; Foroud & Pellis, 2003; Foroud & Whishaw, 2006, 2012; Vasey, Foroud, Duckworth & Kovaxovsky, 2006). The second, Eshkol-Wachmann movement notation (EWMN) (Eshkol & Wachmann, 1958), is the one we have used most extensively (e.g., Kraus, Pellis, & Pellis, 2019; Ottenheimer Carrier, Leca, Pellis & Vasey, 2015; Pellis, 1981b, 1982, 1985; Whishaw & Pellis, 1990), and so is the one that we will outline in this appendix. Prior to proceeding, however, we wish to note that we are not averse to using computerised digitising systems for more precise, quantitative analyses of movement, but we have found that using such systems is more effective once some insight into the organisation of the behaviour is gained from these more qualitative methods (e.g., Melvin et al., 2005; Pellis & Pellis, 1994; Whishaw et al., 2002). It may save you from drowning in a quagmire of data that digitised systems can spit out!

In EWMN, the body is treated as a system of articulated axes (i.e., body and limb segments) (Figure A.1). A limb is any part of a body that either lies between two joints or has a joint and an extremity. These are imagined as straight lines (axes), of constant length, which move with one end fixed to the centre of a sphere (Figure A.2). Movements can then be traced from one location on the sphere to another, giving each position a horizontal and vertical coordinate. As shown in Figure A.2, the locations on the sphere are $45°$ apart, but the unit of angular measurement can be reduced

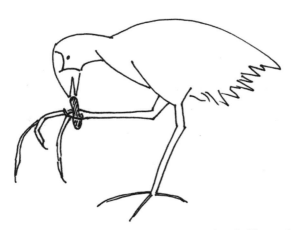

Figure A.1 The drawing shows a swamp hen holding a food item in its foot while manipulating it with its bill. The swamp hen holds the food item in its foot in this elevated position as it manoeuvres its head around so that it can peck, from any orientation, at the pieces of food protruding from its digits. For notation purposes, the limb and body segments can be envisaged as straight lines running along the length of the body parts. For example, a straight line from the tip of the swamp hen's bill to the base of its skull would represent the head, and a line from the ankle to the beginning of its digits would represent the lower leg. Reprinted with permission from Pellis (2011)

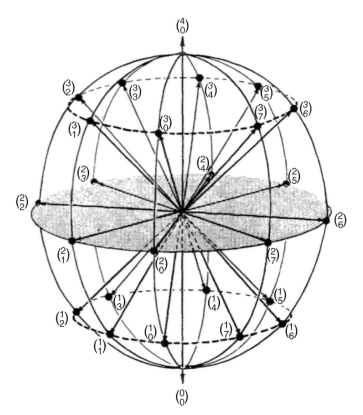

Figure A.2 The EWMN coordinate system. One end of a limb segment is anchored at the centre of the sphere and the other end traces a path of movement on the surface represented as displacements with one unit equals 45°. Each pair of numerals indicates the horizontal coordinate (lower numeral) and the vertical coordinate (upper numeral).
Reproduced from Eshkol & Wachmann (1958)

(e.g., 22.5°) if finer-grain comparisons are needed. The movements of each segment from one location to another can be notated with regard to the shape (planar, conical or rotatory), direction (clockwise or anticlockwise) and magnitude (number of angular units). Designated symbols represent these different facets of movement.

The body is represented on a horizontally ruled page in columns that denote units of time (e.g., frames of a video). The signs for movement are read from left to right and from bottom to top (Figure A.3). As can be seen from the notated page, the contribution of each segment to the overall

Frames	0	1	9	15	20	21	22	25	27	28	29	31		33	34	35	36	38
Head/neck		TF	↑								↓			∩				
Torso				↑														
Right leg:																		
Upper																		
Lower																		
Foot	T																	
Left leg:																		
Upper							↑		↓									
Lower							↑			↓								
Foot	T					T̄	=				∩							TF
Weight					[2]													

Figure A.3 A simplified EWMN score shows a sequence of a swamp hen grasping at a food item that is being held in its bill by its foot. Body elements over time are depicted on the left-hand side and movements over time are on the right / (Vertical columns represent video frames). This particular sequence was observed from the side, and, for simplicity, the magnitude of the movements is not shown. The notation begins with the frame at which the swamp hen picks the food item up with its bill. The various symbols represent movements, with up and down facing arrows indicating vertical movements and the inverted U indicating rotation of the swamp hen's foot and head, respectively. The T and = signs indicate weight-bearing contact and loss of that contact, respectively, while the T with a bar on top indicates non-weight-bearing contact on the ground. TF indicates contact with the food item by the bird's bill and foot, respectively. The row indicating weight denotes shifts of body weight, in this case [2], indicate that the bird shifted its body weight to the right and in doing so, shifted weight away from its left foot. The arcs connecting frames denote the duration of movement. Reprinted with permission from Pellis (2011)

movement and how they are coordinated together is shown. For instance, in Figure A.3, the coordinated leg and body movements of a swamp hen reaching up with its foot to grasp a food item held in its bill are shown. The notation shows that the swamp hen moves the limb that is about to grasp the food item (left foot) to a particular location in space, in front and to the left of its sternum and then holds it there (see lower right section of the notated page). To reach the food item, the swamp hen then makes movements with its head and neck that positions the food near its foot (see upper right section of the notated page). The importance of head movements in enabling the swamp hen to grasp the food item with its foot was revealed by this kind of analysis (Pellis, 2011). For readers interested in exploring this method further, we recommend starting with Golani (1976) and then Eshkol and Wachmann (1958) for the mechanics of how the system works. For present purposes, we will emphasise those aspects of the system that continue to help us overcome our limitations as observers.

EWMN allows the same movements to be notated in several polar coordinate systems. The coordinates of each system are determined with reference to the surrounding, fixed environment, to the subject's own next proximal or distal limb or body segment, or in the case of an interaction with another animal, that animal's body. By transforming the description of the same behaviour from one coordinate system to the next, invariance in the behaviour may emerge in some coordinates but not others. Thus, the behaviour may be invariant in relation to some or all of the following – gravity, the animal's own body, or some aspect of the environment, including another animal in a social encounter (Golani, 1976). As discussed in Chapter 2, an invariant relationship can be a clue as to the perceptions that the animal is maintaining as constant and so guiding its behaviour. The beauty of EWMN is that it enables us to overcome our perceptual bias to notice change and detect invariances that may be critical to understanding the animal's behaviour.

Male greater sage-grouse compete for optimal locations during the mating season to access receptive females. During these interactions, the males perform particular displays to attract females and engage other males agonistically. The males position themselves in what are called 'facing past' encounters and, more rarely, actually strike one another with their wings (Wiley, 1973). In the 'facing past' encounter, the males stand about a body length apart, oriented shoulder-to-shoulder and facing in the opposite direction. This configuration has been interpreted as a display to intimidate one's opponent and was thought to be separate from actual combat. We used videotaped encounters to analyse the movements of both opponents using EWMN (Pellis et al., 2013). For this study, the EWMN score was limited to the horizontal plane, as the birds remained on ground as they manoeuvred around one another, limiting the role of vertical movements.

To keep track of the relationship between the two birds during interactions, the bodies of the two interacting birds were described on three coordinate systems (see Figure A.4). *Partner-wise orientation*: the relationship of the longitudinal axis of one bird relative to the other. One bird is selected as the focal animal and the 45° units are situated in a circle around the longitudinal axis (0–7), with 0 being situated in the direction in which

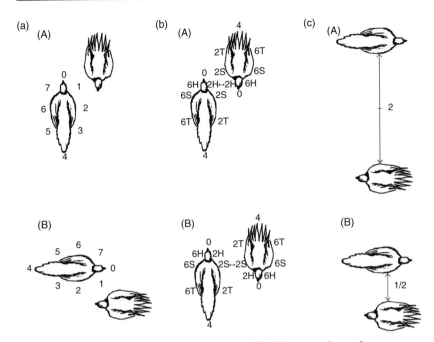

Figure A.4 This figure illustrates the three simplified, horizontal coordinates derived from EWMN that were used to record the inter-animal configuration during interactions. (a) Partner-wise orientation: in this case, even though the sage grouse change their position in space, the relative orientation of their longitudinal axes remains the same. (b) Opposition: In this case, as the birds move, they switch from an opposition that is head-to-head to one that is shoulder-to-shoulder. (c) Relative distance: in this case, as the birds move, they decrease their relative distance to one another. Reprinted with permission from Pellis et al. (2013)

the bird is facing, and continues to be so situated even as the bird changes position in space. The longitudinal axis of the other bird is superimposed over the longitudinal axis of the focal animal and the value pointed at by the anterior of the opponent is given that numerical label. For example, in Figure A.4a, panel A, the focal animal is facing towards the top of the page and the opponent to the bottom of the page, so that if the opponent were imagined to be sitting on top of the focal bird, its bill would be facing past the tail end to value 4, and so would be labelled {4} (with {} signifying coordinates for partner-wise orientation). In panel B, the two birds have moved, with the focal bird rotating around its longitudinal axis by 90° to the right – but so has its opponent. This means that, although both birds have changed their position in space, their relationship to each other remains the same. Since in partner-wise notation, the circle of coordinate values moves with the focal animal, the partner-wise orientation continues to be designated as {4}. *Opposition*: the part of the body of one bird closest to the other is scored. To score this, imagine the EWMN sphere being deflated, so that it is wrapped around the animal's body. The front of the sphere (taking the horizontal value only) would be 0 and this value would be attached to the tip of the bird's beak, and the most posterior point would be 4. Similarly, each side of the bird's body (head, shoulder, torso) would be labelled 2 for the right side and 6 for the left side. In Figure A.4b, panel A, the two birds are shown, with their respective labels attached, standing, in an antiparallel orientation, with their heads in closest proximity to one another. The opposition between the two birds would be designated at 2H/2H, as the sides of their heads would be in closest proximity. In panel B, the birds move forward, relative to one another, which results in their shoulders being in closest proximity (2S/2S). *Relative distance*: given that videotapes were not taken with a measurable frame of reference, the absolute distance in a metric, such as centimetres, was not possible, but the distance in terms of bird lengths (i.e., when the bird is standing upright in a relaxed posture, from the tip of its bill to the base of its tail) could be used to track the relative distance during encounters between the birds. For example, in Figure A.4c, panel A, the two birds are side-by-side but are two body lengths apart. Then, following some movement, in panel B, the two birds have closed the gap between them to about a 0.5 body length. Together, these three coordinates create a space within which the relationship between the two animals can be traced throughout the encounter (Moran et al., 1981; Pellis, 1982).

By using different aspects of the EWMN to record the whole body movements of the individual birds during these encounters, it is possible to determine which of the birds is contributing to changes in their inter-animal relationship (Pellis, 1989). For this study, whole body movements were recorded in two ways. First, changes in the orientation of the longitudinal axis of the body were recorded as rotations in absolute space. For example, in Figure A.4a, the focal animal rotates its body axis 90° to the right, and for this, the sphere (or, in this case, its horizontal dimension) is viewed as remaining stationary as the bird moves within it. If the absolute coordinate in this case involves the front of the sphere being closest to the top of the page, then the bird would be facing (0), and, after rotating, faces (2) (in which () signifies movements in absolute space). In contrast, the opponent in Figure A.4a begins at (4) and ends at (6). Rotation in space is shown as U for clockwise movements, and as an inverted U for anti-clockwise movements. These rotational changes of the body axis in absolute space are called changes in 'Front'.

Second, the movement of the whole body in space can be accomplished by shifting the body's centre of mass in a particular direction, which may or may not involve stepping. These bodily movements were scored – again, only in the horizontal plane – and in a body-wise frame of reference. That is, if the bird leans to its right, it would be designated as [2] and if backwards, [4] (in which [] signifies body-wise notation). This designation would remain the same, irrespective of the bird's position in absolute space; a body shift to the right would be [2], no matter whether the bird was pointing towards (0) or (6) in absolute space. If the shift of body weight involved stepping, then the direction of the shift would be accompanied with the appropriate stepping symbol, S. These shifts in body mass are referred to as changes in 'Weight' (Eshkol & Wachmann, 1958).

The notated movements of the opponents are shown in Figure A.5. At the start of the EWMN score, the positions of the birds are shown, from an aerial perspective, on the far left (under (i)). Starting from the antiparallel configuration with their shoulders opposed and standing at about 1.5 bird lengths apart (i), bird b moves laterally towards bird a, and bird a steps laterally in the opposite direction (as shown by the arrows next to the drawings of the birds). Therefore, at the end of the movements, because b makes a bigger lateral displacement than a, even though the distance between them has closed to one body length, they remain in the same opposition and partner-wise orientation (ii). Bird b then makes another set of movements, stepping obliquely backwards and rotating towards its opponent, but this is coupled with bird a stepping laterally away and rotating towards b.

At the end of the sequence of movements, the birds' orientation in absolute space has changed, as has the inter-animal distance, which has been reduced to half a body length, but their partner-wise orientation and opposition have remained the same (iii). Looking at the arcs designating the duration of the movements of each bird illustrates how the notation

Frames	0	6	7	11	20	23	25	32
a. Front	(4)						U	(3)
Weight			S [2]			S [2]		
Partner wise	{4}							
Opposition	2S/2S							
Distance	1.5		1					0.5
b. Front	(0)					U		(7)
Weight		S [6]				S [5]		
	(i)			(ii)			(iii)	

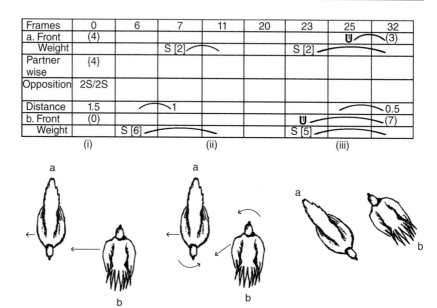

Figure A.5 The top panel shows the notation of a short sequence, around one second, embedded within a fight, in which two clusters of movements by the two opponents are illustrated. The bottom panel shows drawings from the dorsal view of the relative positions of the birds at specific moments in the score. Reprinted with permission from Pellis et al. (2013)

score can be used to identify how the movements between the two birds are correlated and which of these movements lead to changes in the inter-animal configuration. Moreover, the movements by one bird can be seen to counteract the changes that would otherwise be imposed by the movements of the other bird.

If one bird manages to gain a slightly off antiparallel position so that the elbow of its wing facing its opponent is closer, the bird may successfully strike the top of its opponent's head with its elbow. This is the most effective strike, which is able to knock one's opponent off-balance. What the analyses showed was that both birds moved and counter-moved to prevent its opponent from gaining such an advantageous position; this resulted in the birds maintaining the 'facing past' configuration. The 'facing past' encounter is thus not a display separate from combat, but a by-product of the birds manoeuvring to strike one another during combat (Pellis et al., 2013).

Although we have found EWMN of value in gaining insight into the organisation of behaviour, the downside of EWMN is the amount of training involved in gaining proficiency and the limitation in the amount of behaviour that can be notated. Therefore, it is worth pointing out that new methods are emerging that make use of automatic digital image capture and machine learning to track movements of different body parts from the same animal or between animals (e.g., Inayat et al., 2020; Mathis et al, 2018). For example, the kind of trap in configuration between the fighting grouse that we showed using EWMN has also been characterised in a different species and behaviour using such a computerised system (López Pérez et al., 2017).

There is an expanding tool-kit of methods with which to analyse the organisation of behaviour. While these different methods can be complementary, we simply cannot master all of them, so we recommend that readers explore these different methods and then select the ones that are most appealing to them, as long as they bear in mind that different methods may be most useful for different behaviours and different questions. There is no magic bullet.

Appendix B: Practice, Practice, Practice

Vivien's father, George Cosopodiotis, told us that when he was in medical school, his professors stressed the importance of seeing as many conditions as possible, maximising the range of symptoms observed. His stories of students running from one patient to another in search of novel conditions amused us, but the potential knowledge gained from seeing so many patients and symptoms was impressed upon our minds. Thus prepared by George, we were delighted, on beginning work in the laboratory of Philip Teitelbaum, that he encouraged us not just to focus on the behaviours that we were directly studying, but also to see the behaviour being studied by other members of our laboratory and, indeed, in neighbouring laboratories. Having diverse experiences has two important consequences. First, you have a repertoire of examples in your mind against which you can compare and contrast a new phenomenon. For example, after studying play fighting in rats and many other rodents, it was obvious to us that the competition was centered on nuzzling the nape of the neck. But then we came across species that competed for other targets, such as nibbling the cheeks in Syrian hamsters and licking the mouth in Djungarian hamsters (Pellis & Pellis, 2009). Are all these species idiosyncratic in their targets or is there a deeper commonality? As it turned out, what unified all these species, is that the targets competed over during play fighting are derived from adult sexual behaviour. The targets may vary, but the theme remains the same (see Chapters 3 and 5).

Second, exposure to diverse phenomena gets you out of your comfort zone and can sometime lead to deeper insights. For example, after having studied the play fighting of many rodents and primates, we characterised the reciprocity needed for play fighting to remain playful as arising from either partner self-handicapping itself, which allows for role reversals (Pellis & Pellis, 2009). That is, play partners exercise some restraint. Of course, we dismissed reports of the absence of restraint as reflecting the observer's incompetence or not looking closely enough. So, we ourselves then studied the play fighting of a species from a group – pigs – reported not to show restraint during play. Applying techniques that enabled us to track the correlated movements of the partners closely (see Appendix A), we made an important discovery; we were wrong – pigs really do not show restraint (Pellis & Pellis, 2016). This broke the complacency in our own understanding of this phenomenon (play fighting) and made us look at it more closely. While it remains true that for play fighting to remain playful, the partners need to integrate some degree of cooperation to achieve a modicum of reciprocity (see Chapter 5), pulling your punches when attacking and defending is not the only way to achieve this (Pellis & Pellis, 2017). Again, seeing new phenomena can break through observational prejudices accrued from past experiences.

Our recommendation to students and their teachers is to gain as much breadth in your exposure to the diverse behaviours of as wide a range of animals as possible. Similarly, for laboratory-oriented researchers, observe the behaviour of as many different preparations as possible. As zoologists and psychologists, respectively, we arrived in Teitelbaum's laboratory sceptical about the value of studying rats with brain lesions, but he taught us to think of each lesion as producing a different 'species' of rat. Comparing and contrasting across many different types of lesion can yield remarkable insights into the organisation of behaviour (Teitelbaum, 2012). Once we learned this lesson, we added brain damage, drug and hormonal manipulations to our existing repertoire of cross-species and cross-age comparisons in gaining understanding of how a behaviour is organised, as each may modify the way the principles explored in this book (Chapters 2–4) operate. It is those modifications that often allow you to break free from your preconceived ideas.

What diverse experience also highlights is that practice is essential. Watch as many nature documentaries as you can, scour YouTube, sit and watch what your pets do, visit the zoo, all the while trying in your mind's eye to apply the principles we have outlined in this book. Even better, go to the pet shop and buy some crickets, which are usually sold as food for such animals as pet lizards or frogs, and set them up in a terrarium. Exotic, ornamental cockroaches are another option. It does not take much to maintain them at school or at home and with very little manipulation you can create contexts in which they will engage in various behaviours, such as eating, grooming, mating, and fighting (Logue et al., 2009). Tracking the movements of their body parts when engaged in solitary behaviour or correlated movements between partners in social exchanges can be used to identify the perceptions that are maintained as constant (Pellis & Bell, 2020). Simple variations in the way that the enclosure and substrate are structured can reveal the roles of body morphology and context in constraining the movements that shape the behaviour (Pellis, Gray & Cade, 2009). The bottom line: the more you see and the more behaviour you try to analyse, the more adept you will become at describing and understanding animal behaviour, but always look at more in case you get too confident in the veracity of your conclusions.

References

Acharya, L., & Fenton, M. B. (1992). Echolocation behaviour of vespertilionid bats (*Lasiurus cinereus* and *Lasiurus borealis*) attacking airborne targets including arctiid moths. *Canadian Journal of Zoology*, 70, 1292–1298.

Adams, D. B. (1980). Motivational systems of agonistic behavior in muroid rodents: A comparative review and neural model. *Aggressive Behavior*, 6, 295–346.

Adams, N., & Boice, R. (1989). Development of dominance in domestic rats in laboratory and seminatural environments. *Behavioural Processes*, 19, 127–142.

Adler, J. (1975). Chemotaxis in bacteria. *Annual Reviews of Biochemistry*, 44, 341–356.

Alaverdashvili, M., & Whishaw, I. Q. (2013). A behavioral method for identifying recovery and compensation: Hand use in a preclinical stroke model using the single pellet reaching task. *Neuroscience & Biobehavioral Reviews*, 37, 950–967.

Alaverdashvili, M., Leblond, H., Rossignol, S., & Whishaw, I. Q. (2008). Cineradiographic (video X-ray) analysis of skilled reaching in a single pellet reaching task provides insight into relative contribution of body, head, oral, and forelimb movements in rats. *Behavioural Brain Research*, 192, 232–247.

Alberts, J. R. (2005). Infancy. In I. Q. Whishaw & B. Kolb (Eds.), *The behavior of the laboratory rat: A handbook with tests* (pp. 266–277). New York: Oxford University Press.

Alberts, J. R. (2007). Huddling by rat pups: Ontogeny of individual and group behavior. *Developmental Psychobiology*, 49, 22–32.

Alberts, J. R. (2012). Observe, simplify, titrate model, and synthesize: A paradigm for analyzing behavior. *Behavioural Brain Research*, 231, 250–261.

Alberts, J. R., & Galef, B. G. (1973). Olfactory cues and movement: Stimuli mediating intraspecific aggression in the wild Norway rat. *Journal of Comparative & Physiological Psychology*, 85, 233–242.

Alcock, J. (2013). *Animal behavior: An evolutionary approach.* 10th edition. Oxford: Sinauer Associates.

Aldis, O. (1975). *Play fighting.* New York: Academic Press.

Almli, R. C., & Fisher, R. S. (1977). Infant rats: Sensorimotor ontogeny and effects of substantia nigra destruction. *Brain Research Bulletin, 2,* 425–459.

Altman, J., & Sudarshan, K. (1975). Postnatal development of locomotion in the laboratory rat. *Animal Behaviour, 23,* 896–920.

Altmann, J. (1974). Observational study of behaviour: Sampling methods. *Behaviour, 48,* 1–41.

Anderson, D. J., & Perona, P. (2014). Toward a science of computational ethology. *Neuron, 84,* 18–31.

Archer, J. (1988). *The behavioural biology of aggression.* Cambridge: Cambridge University Press.

Archer, J., & Huntingford, F. (1994). Game theory models and escalation of animal fights. In M. Potegal & J. F. Knutson (Eds.), *Dynamics of aggression: Biological and social processes in dyads and groups* (pp. 3–31). Hillsdale, NJ: Lawrence Erlbaum Associates.

Armitage, J. P. (1997). Behavioural responses of bacteria to light and oxygen. *Archives of Microbiology, 168,* 249–261.

Baerends, G. P. (1976). The functional organization of behaviour. *Animal Behaviour, 24,* 726–738.

Barnett, S. A. (1975). *The rat: A study in behavior.* Chicago: The University of Chicago Press.

Barnett, S. A., & Marples, T. G. (1981). The 'threat posture' of wild rats: A social signal or an anthropomorphic assumption? In P. F. Brain & D. Benton (Eds.), *Multidisciplinary approaches to aggression research* (pp. 39–52). Amsterdam, the Netherlands: Elsevier/North-Holland Biomedical Press.

Barrett, L. (2011). *Beyond the brain: How body and environment shape animal and human minds.* Princeton: Princeton University Press.

Bell, H. C. (2013). Control in living systems: An exploration of the cybernetic properties of interactive behaviour. Unpublished doctoral dissertation, University of Lethbridge, Lethbridge, AB.

Bell, H. C. (2014). Behavioral variability in the service of constancy. *International Journal of Comparative Psychology, 27,* 196–217.

Bell, H. C., & Pellis, S. M. (2011). A cybernetic perspective on food protection in rats: Simple rules can generate complex and adaptable behaviour. *Animal Behaviour, 82,* 659–666.

Bell, H. C., Johnson, E., Judge, K. A., Cade, W. H., & Pellis, S. M. (2012). How is a cricket like a rat? Insights from the application of cybernetics to evasive food protective behaviour. *Animal Behaviour, 84,* 843–851.

Bell, H. C., Bell, G. D., Schank, J. A., & Pellis, S. M. (2015). Attack and defense of body targets in play fighting: Insights from simulating the 'keep away game' in rats. *Adaptive Behaviour*, 23, 371–380.

Bernard, C. (1865/1927). *An introduction to the study of experimental medicine* (H. C. Green, trans.). New York: Macmillan.

Berridge, K. C., & Fentress, J. C. (1986). Deterministic versus probabilistic models of behaviour: Taste-elicited actions in rats as a case study. *Animal Behaviour*, 34, 871–880.

Berthoz, A. (2009). The human brain 'projects' upon the world, simplifying principles and rules for perception. In A. Berthoz & Y. Christen (Eds.), *Neurobiology of the 'Umwelt': How living beings perceive the world* (pp. 17–27). Berlin, Germany: Springer-Verlag.

Berthoz, A., & Christen, Y. (2009). *Neurobiology of the 'Umwelt': How living beings perceive the world*. Berlin, Germany: Springer-Verlag.

Blake, B. E., & McCoy, K. A. (2015). Hormonal programming of rat social play behavior: Standardized techniques will aid synthesis and translation to human health. *Neuroscience & Biobehavioral Reviews*, 55, 184–197.

Blanchard, D. C., & Blanchard, R. J. (1990). The colony model of aggression and defense. In D. A. Dewsbury (Ed.), *Contemporary issues in comparative psychology* (pp. 410–430). Sunderland, MA: Sinauer Associates.

Blanchard, R. J., & Blanchard, D. C. (1994). Environmental targets and sensorimotor systems in aggression and defence. In S. J. Cooper & C. A. Hendrie (Eds.), *Ethology and Psychopharmacology* (pp. 133–157). New York: John Wiley & Sons.

Blanchard, R. J., Blanchard, D. C., Takahashi, T., & Kelley, M. J. (1977). Attack and defensive behaviour in the albino rat. *Animal Behaviour*, 25, 622–634.

Blanchard, R. J., O'Connell, V., & Blanchard, D. C. (1979). Attack and Defensive behaviors in the albino mouse. *Aggressive Behavior*, 5, 622–634.

Bolhuis, J. J., & Giraldeau, L-A. (2005). *The behavior of animals: Mechanisms, function, and evolution*. Oxford: Blackwell.

Bowers, R. I. (2017). Behavior systems. In J. Vonk & T. K. Shackelford (Eds.), *Encyclopedia of animal cognition and behavior* (pp. 1–8). Berlin, Germany: Springer.

Brain, P. F. (1981). Differentiating types of attack and defense in rodents. In P. F. Brain & D. Benton (Eds.), *Multidisciplinary approaches to aggression research* (pp. 53–77). Amsterdam, the Netherlands: Elsevier/North-Holland Biomedical Press.

Brain, P. F., Parmigiani, S., Blanchard, R., & Mainardi, D. (1990). *Fear and Defense*. London: Harwood Academic Publishers.

Breed, M. D., Meaney, C., Deuth, D., & Bell, W. J. (1981). Agonistic interactions of two cockroach species, *Gromphadorhina portentosa* and *Supella longipalpa* (Orthoptera (Dictyoptera): Blaberidae, Blattellidae). *Journal of the Kansas Entomological Society*, 54, 197–208.

Brentari, C. (2015). *Jakob von Uexküll: The discovery of the Umwelt between biosemiotics and theoretical biology*. Berlin, Germany: Springer-Verlag.

Brindley, G. S. (1965). How does an animal that is dropped in a non-upright posture know the angle through which it must turn in the air so that its feet point to the ground? *Journal of Physiology* (London), 180, 20.

Brown, A. E. X., & de Bivort, B. (2018). Ethology as a physical science. *Nature Physics*, https://doi.org/10.1038/s41567-018-0093-0

Brown, A. R., & Teskey, G. C. (2014). Motor cortex is functionally organized as a set of spatially distinct representations for complex movements. *The Journal of Neuroscience*, 34, 13574–13585.

Brown, R. E. (1985). The rodents II: Suborder Myomorpha. In R. E. Brown & D. W. MacDonald (Eds.), *Social odours in mammals*, Vol. 1. (pp. 345–457). Oxford: Clarendon Press.

Burgdorf, J., Kroes, R. A., Moskal, J. R., Pfaus, J. G., Brudzynski, S. M., & Panksepp, J. (2008). Ultrasonic vocalizations of rats (*Rattus norvegicus*) during mating, play, and aggression: Behavioral concomitants, relationship to reward, and self-administration of playback. *Journal of Comparative Psychology*, 122, 357–367.

Burghardt, G. M. (1980). Behavioral and stimulus correlates of vomeronasal functioning in reptiles: Feeding, grouping, sex, and tongue use. In D. Müller-Schwarze & R. M. Silverstein (Eds.), *Chemical signals. Vertebrates and aquatic invertebrates* (pp. 275–301). New York, NY: Plenum Press.

Burghardt, G. M. (2005). *The genesis of animal play: Testing the limits*. Cambridge: MIT Press.

Burghardt, G. M., & Bowers, R. I. (2017). From instinct to behavior systems: An integrated approach to ethological psychology. In J. Call (Ed.-in-Chief), *APA handbook of comparative psychology: Vol. 1. Basic concepts, methods, neural substrate, and behavior* (pp. 333–364). Washington, DC: American Psychological Association.

Burghardt, G. M., Bartmess-LeVasseur, J. N., Browning, S. A., Morrison, K. E., Stec, C. L., Zachau, C. E., & Freeberg, T. M. (2012). Minimizing observer bias in behavioral studies: A review and recommendations. *Ethology*, 118, 511–517.

Burke, C. J., Kisko, T. M., Euston, D. R., & Pellis, S. M. (2018). Do juvenile rats use specific ultrasonic calls to coordinate their social play? *Animal Behaviour*, 140, 81–92.

Burkhart, R. W. Jr. (2005). *Patterns of behavior. Konrad Lorenz, Niko Tinbergen, and the founding of ethology.* Chicago: The University of Chicago Press.

Carter, S. C. (1985). Female sexual behavior. In H. I. Siegel (Ed.), *The hamster: Reproduction and behavior* (pp. 173–189). New York: Plenum Press.

Casarrubea, M., Magnusson, M. S., Anguera, M. T., Jonsson, G. K., Castañer, M., Santangelo, A., Palacino, M., Aiello, S., Faulisi, F., Raso, G., Puigarnau, S., Camerino, O., di Giovanni, G., & Crescimanno, G. (2018). T-pattern detection and analysis for the discovery of hidden features of behaviour. *Journal of Neuroscience Methods,* 310, 24–32.

Cheng, J.-T., Schallert, T., DeRyck, M., & Teitelbaum, P. (1981). Galloping induced by pontine tegmentum damage in rats: A form of "Parkinsonian festination" not blocked by haloperidol. *Proceedings of the National Academy of Science* (USA), 78, 3279–3283.

Clark, A. (1996). *Being there: Putting brain, body and world together again.* Cambridge, MA: MIT Press.

Clark, D. C., & Moore, A. J. (1994). Social interactions and aggression among male Madagascar hissing cockroaches (*Gromphadorhina portentosa*) in groups (Dictyoptera: Blaberidae). *Journal of Insect Behavior,* 7, 199–215.

Clark, R. W. (2016). The hunting and feeding behavior of wild rattlesnakes. In G. W. Schuett, R. S. Reiserer, C. F. Smith & M. J. Feldner (Eds.), *The rattlesnakes of Arizona* (pp. 91–118). Rodeo, NM: Eco Publishing.

Colvin, P. V. (1973). Agonistic behaviour in males of five species of voles. *Animal Behaviour,* 21, 471–480.

Cools, A. R., Brachten, R., Heeren, D., Willemen, A., & Ellenbroek, B. (1990). Search for the neurobiological profile of individual-specific features of Wistar rats. *Brain Research Bulletin,* 24, 49–69.

Cooper, W. E., Jr., & Burghardt, G M. (1990). Vomerolfaction and vomodor. *Journal of Chemical Ecology,* 16, 103–105.

Cowan, P. E. (1981). Early growth and development of roof rats. *Journal of Mammalogy,* 45, 239–250.

Cziko, G. (2000). *The things we do. Using the lessons of Bernard and Darwin to understand the what, how, and why of our behavior.* Cambridge, MA: MIT Press.

Dawkins, M. S. (2007). *Observing animal behaviour. Design and analysis of quantitative data.* Oxford: Oxford University Press.

Delius, J. D. (1969). A stochastic analysis of the maintenance behaviour of skylarks. *Behaviour,* 33, 137–177.

Dempster, E. R., & Perrin, M. R. (1989). A comparative study of agonistic behaviour in hairy-footed gerbils (genus *Gerbillurus*). *Ethology,* 83, 43–59.

Donaldson, T. N., Barto, D., Bird, C. W., Magcalas, C. M., Rodriguez, C. I., Fink, B. C., & Hamilton, D. A. (2018). Social order: Using the sequential structure of social interaction to discriminate abnormal social behavior in the rat. *Learning & Motivation*, 61, 41–51.

Dorwood, D. F. (1977). A case of comeback: The Cape Barren goose. *Australian Natural History*, 19, 130–135.

Drai, D., & Golani, I. (2001). See: A tool for the visualization and analysis of rodent exploratory behavior. *Neuroscience & Biobehavioral Reviews*, 25, 409–426.

Driver, P. M., & Humphries, N. (1988). *Protean behavior: The biology of unpredictability*. Oxford: Oxford University Press.

Dugatkin, L. A., & Trut, L. (2017). *How to tame a fox (and build a dog)*. Chicago: University of Chicago Press.

Eberhard, W. G. (1990). Imprecision in the behavior of *Leptomorphus* sp. (Diptera, Mycetophilidae) and evolutionary origin of new behavior patterns. *Journal of Insect Behavior*, 3, 327–357.

Eilam, D. (1997). Postnatal development of body architecture and gait in several rodent species. *Journal of Experimental Biology*, 200, 1339–1350.

Eilam, D., & Golani, I. (1988). The ontogeny of exploratory behavior in the house rat (*Rattus rattus*): The mobility gradient. *Developmental Psychobiology*, 21, 679–710.

Eilam, D., & Golani, I. (1989). Home base behavior of rats (*Rattus norvegicus*) exploring a novel environment. *Behavioural Brain Research*, 34, 199–211.

Eilam, D., & Golani, I. (1990). Home base behavior in amphetamine-treated tame wild rats (*Rattus norvegicus*). *Behavioural Brain Research*, 36, 161–170.

Eilam, D., Adijes, M., & Velinsky, J. (1995). Uphill locomotion in mole rats: A possible advantage of backward locomotion. *Physiology & Behavior*, 58, 483–489.

Ellis, M. E. (1982). Evolution of aversive information processing: A temporal trade-off hypothesis. *Brain, Behavior & Evolution*, 21, 151–160.

Eshkol, N, & Wachmann, A. (1958). *Movement notation*. London: Weidenfeld & Nicholson.

Ewert, J-P. (2005). Stimulus perception. In J. J. Bolhuis & Giraldeau, L-A. (Eds.), *The behavior of animals. Mechanisms, function, and evolution* (pp. 13–40). Oxford: Blackwell.

Fagen, R. A. (1981). *Animal play behavior*. New York: Oxford University Press.

Fagen, R., Conitz, J., & Kunibe, E. (2000). Observing behavioral qualities. *International Journal of Comparative Psychology*, 10, 167–179.

Fernández-Espejo, E., & Mir, D. (1990). Ethological analysis of the male rat's socioagonistic behaviour in a resident-intruder paradigm. *Aggressive Behavior*, 16, 41–55.

Field, E. F., & Pellis, S. M. (2008). The brain as the engine of sex differences in the organization of movement in rats. *Archives of Sexual Behavior*, 37, 30–42.

Field, E. F., & Whishaw, I. Q. (2008). Sex differences in the organization of movement. In J. B. Becker, K. J. Berkley, N. Geary, E. Hampson, J. P. Herman, & E. A. Young (Eds.), *Sex differences in the brain from genes to behavior* (pp.155–175). New York: Oxford University Press.

Field, E. F., Whishaw, I. Q., & Pellis, S. M. (1996). An analysis of sex differences in the movement patterns used during the food wrenching and dodging paradigm. *Journal of Comparative Psychology*, 110, 298–306.

Field, E. F., Martens, D. J., Watson, N. V., & Pellis, S. M. (2005). Sex differences in righting from supine to prone: A masculinized skeletomusculature is not required. *Journal of Comparative Psychology*, 119, 238–245.

Finkelstein, L. (1982). What is not measurable, make measurable. *Measurement & Control*, 15, 25–32.

Flash, T., & Hochner, R. (2005). Motor primitives in vertebrates and invertebrates. *Current Opinion in Neurobiology*, 15, 660–665.

Foroud, A., & Pellis, S. M. (2003). The development of 'roughness' in the play fighting of rats: A Laban Movement Analysis perspective. *Developmental Psychobiology*, 42, 35–43.

Foroud, A., & Whishaw, I. Q. (2006). Changes in the kinematic structure and non-kinematic features of movements during skilled reaching after stroke: A Laban Movement Analysis in two case studies. *Journal of Neuroscience Methods*, 158, 137–149.

Foroud, A., & Whishaw, I.Q. (2012). The consummatory origins of visually guided reaching in human infants: A dynamic integration of whole-body and upper-limb movements. *Behavioural Brain Research*, 231, 343–355.

Galef, B. G. Jr. (1996). Social enhancement of food preferences in Norway rats: A brief review. In C. M. Heyes & B. G. Galef, Jr. (Eds.), *Social learning in animals: The roots of culture*, (pp. 49–64). San Diego, CA: Academic Press.

Gallistel, C. R. (1980). *The organization of action: A new synthesis*. Hillsdale, NJ: Lawrence Erlbaum Associates.

Gambaryan, P. P. (1974). *How mammals run: Anatomical adaptations*. New York: John Wiley & Sons.

Garamszegi, L. Z., Calhim, S., Gergely, N. D., Hurd, P. L., Jørgensen, C., Kutsukake, N., Lajeunesse, M. J., Pollard, K. A., Schielzeth, H.,

Symonds, M. R. E., & Nakagawa, S. (2009). Changing philosophies and tools for statistical inferences in behavioral ecology. *Behavioral Ecology*, 20, 1363–1375.

Geist, V. (1978). On weapons, combat and ecology. In L. Krames, P. Pliner & T. Alloway (Eds.), *Advances in the study of communication and affect. Vol. 4, Aggression, dominance and individual spacing* (pp. 1–30). New York: Plenum Press.

Glimcher, P. W. (2003). *Decisions, uncertainty, and the brain: The science of neuroeconomics.* Cambridge, MA: MIT Press.

Golan, L., Radcliffe, C., Miller, T., O'Connell, B., & Chiszar, D. (1982). Trailing behavior in prairie rattlesnakes (*Crotalus viridus*). *Journal of Hepertology*, 16, 287–293.

Golani, I. (1976). Homeostatic motor processes in mammalian interactions: A choreography of display. In P. P. G. Bateson & P. H. Klopfer (Eds.), *Perspectives in ethology*, vol. 2. (pp. 69–134). New York: Plenum.

Golani, I. (1981). The search for invariants in behavior. In K. Immelmann, G. W. Barlow, L. Petrinovich & M. Cain (Eds.), *Behavioral development: The Bielefeld interdisciplinary project* (pp. 372–390). Cambridge: Cambridge University Press.

Golani, I. (2012). The developmental dynamics of behavioral growth processes in rodent egocentric and allocentric space. *Behavioural Brain Research*, 231, 309–316.

Golani, I., Wolgin, D. L., & Teitelbaum, P. (1979). A proposed natural geometry of recovery from akinesia in the lateral hypothalamic rat. *Brain Research*, 164, 237–267.

Gomez-Marin, A., & Ghazanfar, A. A. (2019). The life of behavior. *Neuron*, 104, 25–36.

Gomez-Marin, A., Partoune, N., Stephens, G. J., & Louis, M. (2012). Automated tracking of animal posture and movement during exploration and sensory orientation behaviors. *PLoS ONE*, 7(8): e41642. doi:10.1371/journal.pone.0041642

Gomez-Marin, A., Paton, J. J., Kampff, A. R., Costa, R. M., & Mainen, Z. F. (2014). Big behavioral data: Psychology, ethology and the foundations of neuroscience. *Nature Neuroscience*, 17, 1455–1462.

Graziano, M. S. A. (2009). *The intelligent movement machine. An ethological perspective on the primate motor system.* New York: Oxford University Press.

Graziano, M. S. A. (2016). Ethological action maps: A paradigm shift for the motor cortex. *Trends in Cognitive Sciences*, 20, 121–132.

Grillner, S. (1975). Locomotion in vertebrates: Central mechanisms and reflex interaction. *Physiological Reviews*, 55, 247–304.

Guerra, P. A., & Mason, A. C. (2005). Information on resource quality mediates aggression between male Madagascar hissing cockroaches, *Gromphadorhina portentosa* (Dictyoptera: Blaberidae). *Ethology*, 111, 626–637.

Han, X., Luo, S., & Han, S. (2016). Embodied neural responses to others' suffering. *Cognitive Neuroscience*, 7, 114–127.

Hartstone-Rose, A., Dickinson, E., Boettecher, M. L., & Herrel, A. (2019). A primate with a Panda's thumb: The anatomy of the pseudothumb of *Daubentonia madagascariensis*. *American Journal of Physical Anthropology*, 171, 8–16.

Hempel, C. G., & Oppenheim, P. (1948). Studies in the logic of explanation. *Philosophy of Science*, 15, 135–175.

Herzfeld, D. J., & Shadmehr, R. (2014). Motor variability is not noise, but grist for the learning mill. *Nature Neuroscience*, 17, 149–150.

Himmler, B. T., Pellis, V. C., & Pellis, S. M. (2013). Peering into the dynamics of social interactions: Measuring play fighting in rats. *Journal of Visualized Experiments*, 71, e4288.

Himmler, B. T., Kisko, T. M., Euston, D. R., Kolb, B., & Pellis, S. M. (2014). Are 50-kHz calls used as play signals in the playful interactions of rats? I. Evidence from the timing and context of their use. *Behavioural Processes*, 106, 60–66

Himmler, B. T., Stryjek, R., Modlińska, K., Derksen, S. M., Pisula, W., & Pellis, S. M. (2013). How domestication modulates play behavior: A comparative analysis between wild rats and a laboratory strain of *Rattus norvegicus*. *Journal of Comparative Psychology*, 127, 453–464.

Himmler, S. M., Lewis, J. M., & Pellis, S. M. (2014). The development of strain typical defensive patterns in the play fighting of laboratory rats. *International Journal of Comparative Psychology*, 27, 385–396.

Himmler, S. M., Himmler, B. T., Pellis, V. C., & Pellis, S. M. (2016). Play, variation in play and the development of socially competent rats. *Behaviour*, 153, 1103–1137.

Himmler, S. M., Himmler, B. T., Stryjek, R., Modlińska, K., Pisula, W., & Pellis, S. M. (2016). Pinning in the play fighting of rats: A comparative perspective with some methodological recommendations. *International Journal of Comparative Psychology*, 29, 1–14.

Himmler, S. M., Modlińska, K., Stryjek, R., Himmler, B. T., Pisula, W., & Pellis, S. M. (2014). Domestication and diversification: A comparative analysis of the play fighting of the brown Norway, Sprague-Dawley, and Wistar strains of laboratory rats. *Journal of Comparative Psychology*, 128, 318–327.

Hinde, R. A. (1982). *Ethology: Its nature and relationship with other sciences.* Oxford: Oxford University Press.

Hogan, J. A. (2001). Development of behavior systems. In E. M. Blass (Ed.), *Handbook of behavioral neurobiology (Vol. 13). Developmental psychobiology* (pp. 229–279). Berlin, Germany: Springer.

Hole, G. J., & Einon, D. F. (1984). Play in rodents. In P. K. Smith (Ed.), *Play in animals and children* (pp. 95–117). Oxford: Blackwell.

Hood, B. M. (2010). *The science of superstition.* New York: Harper Collins Publishers.

Horrobin, D. F. (1970). *Principles of biological control.* Aylesbury, UK: Medical and Technical Publishing Co., Ltd.

Horwich, R. H. (1972). The ontogeny of social behavior in the gray squirrel (*Sciurus carolinensis*). *Zeitschrift für Tierpsychologie*, Supplement no. 8, 1–103.

Hough, R. (2001). *The final confession of Mabel Stark.* Toronto, ON: Random House Canada.

Hunsperger, R. W. (1983). A neuroethological study of sexual and predatory aggression in the domestic cat. In J.-P. Ewert, R. R. Capranica, & D. J. Ingle (Eds.), *Advances in vertebrate neuroethology* (pp. 1151–1166). New York: Plenum Press.

Huntingford, F. A., & Turner, A. K. (1987). *Animal conflict.* London: Chapman & Hall.

Hurst, J. L., Barnard, C. J., Hare, R., Wheeldon, E. B., & West, C. D., (1996). Housing and welfare in laboratory rats: Time-budgeting and pathophysiology in single sex groups. *Animal Behaviour, 52*, 335–360.

Hutchinson, A. (1977). *Labanotation: The system of analyzing and recording movement.* 3rd edition. New York: Theatre Arts Books.

Huxley, T. H. (1901). *The scientific memoirs of Thomas Henry Huxley*, Vol. 3. London: Macmillan.

Inayat, S., Singh, S., Ghasroddaashti, A., Qandeel, Egodage, P, Whishaw, I. Q., & Mohajerani, M. H. (2020). A Matlab-based toolbox for characterizing behavior of rodents engaged in string-pulling. *eLife*, 9, e54540

Ivanco, T. L., Pellis, S. M., & Whishaw, I. Q. (1996). Skilled movements in prey catching and in reaching by rats (*Rattus norvegicus*) and opossums (*Monodelphis domestica*): Relations to anatomical differences in motor systems. *Behavioural Brain Research, 79*, 163–182.

Jannett, F. J. Jr. (1981). Scent mediation of intraspecific, interspecific, and intergeneric agonistic behavior among sympatric species of voles (Microtinae). *Behavior, Ecology & Sociobiology*, 9, 273–296.

Johnsgard, P. A. (1965). *Handbook of waterfowl behavior.* New York: Comstock Publishing Associates.

Jusufi, A., Zeng, Y., Full, R. J., & Dudley, R. (2011). Aerial righting reflexes in flightless animals. *Integrative & Comparative Biology*, 51, 937–943.

Kaas, J. H., Gharbawie, O. A., & Stepniewska, I. (2013). Cortical networks for ethologically relevant behaviors in primates. *American Journal of Primatology*, 75, 407–414.

Kline, R. B. (2013). *Beyond significance testing: Statistics reform in the behavioral sciences.* 2nd edition. Washington, DC: American Psychological Association.

Kolb, B., & Whishaw, I. Q. (2015). *Fundamentals of human neuropsychology.* 5th edition. New York: Worth Publishing.

Kolb, B., Whishaw, I. Q., & Teskey, G. C. (2019). *An introduction to brain and behavior.* 6th edition. New York: Worth Publishing.

Kraus, K. L., Pellis, V. C., & Pellis, S. M. (2019). Targets, tactics and cooperation in the play fighting of two genera of Old World Monkeys (*Mandrillus* and *Papio*): Accounting for similarities and differences. *International Journal of Comparative Psychology*, 32, 1–25.

Lakke, J. P. W. P. (1985). Axial apraxia in Parkinson's disease. *Journal of Neurological Science*, 69, 37–46.

Lehner, P. N. (1996). *Handbook of ethological methods.* 2nd edition. Cambridge: Cambridge University Press.

Lelard, T., Jamon, M., Gasc, J.-P., & Vidal, P.P. (2006). Postural development in rats. *Experimental Neurology*, 202, 112–124.

Leonelli, S. (2019). The challenges of big data biology. *eLife*, 8, e47381

Lerwill, C. J., & Makings, P. (1971). The agonistic of behaviour of the golden hamster *Mesocricetus auratus* (Waterhouse). *Animal Behaviour*, 19, 714–721.

Leyhausen, P. (1979). *Cat behavior. The predatory and social behavior of domestic and wild cats.* New York: Garland STPM Press.

Lind, H. (1959). The activation of an instinct caused by a "transitional action". *Behaviour*, 14, 123–135.

Lloyd, G. E. R. (1968). *Aristotle: The growth and structure of his thought.* Cambridge: Cambridge University Press.

Loewen, I., Wallace, D. G., & Whishaw, I. Q. (2005). The development of spatial capacity in piloting and dead reckoning by infant rats: Use of the huddle as a home base for spatial navigation. *Development Psychobiology*, 46, 350–361.

Logue, D. M., Mishra, S., McCaffrey, D., Ball, D., & Cade, W. H. (2009). A behavioral syndrome linking courtship behavior toward males and females predicts reproductive success from a single mating in the hissing cockroach, *Gromphadorhina portentosa*. *Behavioral Ecology*, 20, 781–788.

López Pérez, D., Leonardi, G., Niedźwiecka, A., Radkowska, A., Rączaszek-Leonardi, J., & Tomalski, P. (2017). Combining recurrence analysis and automatic movement extraction from video recordings to study behavioral coupling in face-to-face parent-child interactions. *Frontiers in Psychology*, 8. doi:10.3389/fpsyg.2017.02228

Lorenz, K. Z. (1973). The fashionable fallacy of dispensing with description. *Naturewissenschaften*, 60, 1–19.

Lyon, P. (2007). From quorum to cooperation: Lessons from bacterial sociality for evolutionary theory. *Studies in the History & Philosophy of Science Part C: Studies in the History and Philosophy of Biology & Biomedical Sciences*, 38, 820–833.

MacDonnell, M. F., & Flynn, J. P. (1966). Sensory control of hypothalamic attack. *Animal Behaviour*, 14, 399–405.

MacNulty, D. R., Mech, L. D., & Smith, D. W. (2007). A proposed ethogram of large-carnivore predatory behavior, exemplified by the wolf. *Journal of Mammalogy*, 88, 595–605.

Magnus, R. (1924). *Korperstellung*. Berlin, Germany: Springer.

Magnus, R. (1926). On the co-operation and interference of reflexes from other sense organs with those of the labyrinths. *Laryngoscope*, 36, 701–713.

Marken, R. S. (2002). Looking at behavior through control theory glasses. *Review of General Psychology*, 6, 260–270.

Markus, E. J., & Petit, T. L. (1987). Neocortical synaptogenesis, aging and behavior: Lifespan development in the motor-sensory system of the rat. *Experimental Neurology*, 96, 262–279.

Martens, D. J., Whishaw, I. Q., Miklyaeva, E. I., & Pellis, S. M. (1996). Spatio-temporal impairments in limb and body movements during righting in an hemiparkinsonian rat analogue: Relevance to axial apraxia in humans. *Brain Research*, 733, 253–262.

Martin, P., & Bateson, P. (2007). *Measuring behaviour. An introductory guide.* 3rd edition. Cambridge: Cambridge University Press.

Mathis, A., Mamidanna, P., Cury, K. M., Abe, T., Murthy, V. N., Mathis, M. W., & Bethge, M. (2018). DeepLabCut: Markerless pose estimation of user-defined body parts with deep learning. *Nature Neuroscience*, 10.1038/s41593-018-0209-y

May, C. J., & Schank, J. C. (2009). The interaction of body morphology, directional kinematics and environmental structure in the generation of neonatal rat (*Rattus norvegicus*) locomotor behavior. *Ecological Psychology*, 21, 308–333.

May, C. J., Schank, J. C., & Joshi, S. (2011). Modeling the influence of morphology on the movement ecology of groups of infant rats *(Rattus norvegicus). Adaptive Behavior*, 19, 280-291.

May, C. J., Schank, J. G., Joshi, S., Tran, J., Taylor, R. J., & Scott, I.-E. (2006). Rat pups and random robots generate similar self-organized and intentional behavior. *Complexity*, 12, 53–66.

McFarland, D. J. (1971). *Feedback mechanisms in animal behaviour*. London: Academic Press.

Melvin, K. G., Doan, J., Pellis, S. M., Brown, L., Whishaw, I. Q., & Suchowersky, O. (2005). Pallidal deep brain stimulation and L-dopa do not improve qualitative aspects of skilled reaching in Parkinson's disease. *Behavioural Brain Research*, 160, 188–194.

Meredith, M., & Burghardt, G. M. (1978). Electrophysiological studies of the tongue and accessory olfactory bulb in garter snakes. *Physiology & Behavior*, 21, 1001–1008.

Miller, E. H. (1975). Walrus ethology. I. The social role of tusks and applications of multidimensional scaling. *Canadian Journal of Zoology*, 53, 590–613.

Miller, G. F. (1997). Protean primates: The evolution of adaptive unpredictability in competition and courtship. In A. Whiten & R. W. Byrne (Eds.), *Machiavellian Intelligence II: Extensions and evaluations* (pp. 312–340). Cambridge: Cambridge University Press.

Mook, D. G. (1996). *Motivation. The organization of action*. 2nd edition. New York: W. W. Norton & Company.

Moran, G., Fentress, J. C., & Golani, I. (1981). A description of relational patterns of movement during 'ritualized fighting' in wolves. *Animal Behaviour*, 29, 1146–1165.

Muir, G. D. (2000). Early ontogeny of locomotor behaviour: A comparison between altricial and precocial animals. *Brain Research Bulletin*, 53, 719–726.

Muro, C., Escobedo, R., Spector, L., & Coppinger, R. (2011). Wolf-pack (*Canis lupus*) hunting strategies emerge from simple rules in computational simulations. *Behavioral Processes*, 88, 192–197.

Nelson, J., & Gemmell, R. (2004). Implications of marsupial births for an understanding of behavioural development. *International Journal of Comparative Psychology*, 17, 53–70.

Nevitt, G. A., Losekoot, M., & Weimerskirch, H. (2008). Evidence for olfactory search in wandering albatross, *Diomedea exulans*. *Proceedings of the National Academy of Sciences* (USA), 105, 4576–4581.

Niedenthal, P. M. (2007). Embodying emotion. *Science*, 316, 1002–1005.

Ottenheimer Carrier, L., Leca, J. B., Pellis, S. M., & Vasey, P. L. (2015). A structural comparison of female–male and female–female mounting in Japanese macaques (*Macaca fuscata*). *Behavioural Processes*, 119, 70–75.

Palagi, E., Cordoni, G., Demuru, E., & Bekoff, M. (2016). Fair play and its connection with social tolerance, reciprocity and the ethology of peace. *Behaviour*, 153, 1195–1216.

Panksepp, J. (1981). The ontogeny of play in rats. *Developmental Psychobiology*, 14, 327–332.

Panksepp, J. (1998). *Affective neuroscience*. New York: Oxford University Press.

Panksepp, J., Normansell, L., Cox, J. F., & Siviy, S. M. (1994). Effects of neonatal decortication on the social play of juvenile rats. *Physiology & Behavior*, 56, 429–443.

Park, R. (2000). *Voodoo science: The road from foolishness to fraud*. Oxford: Oxford University Press.

Pellis, S. M. (1981a). Exploration and play in the behavioural development of the Australian magpie *Gymnorhina tibicen*. *Bird Behaviour*, 3, 37–49.

Pellis, S. M. (1981b). A description of social play by the Australian magpie *Gymnorhina tibicen* based on Eshkol-Wachman notation. *Bird Behaviour*, 3, 61–79.

Pellis, S. M. (1982). An analysis of courtship and mating in the Cape Barren goose *Cereopsis novaehollandiae* Latham based on Eshkol-Wachman Movement Notation. *Bird Behaviour*, 4, 30–41.

Pellis, S. M. (1985). What is "fixed" in a Fixed Action Pattern? A problem of methodology. *Bird Behaviour*, 6, 10–15.

Pellis, S. M. (1988). Agonistic versus amicable targets of attack and defense: Consequences for the origin, function and descriptive classification of play-fighting. *Aggressive Behavior*, 14, 85–104.

Pellis, S. M. (1989). Fighting: The problem of selecting appropriate behavior patterns. In R. J. Blanchard, P. F. Brain, D. C. Blanchard & S. Parmigiani (Eds.), *Ethoexperimental approaches to the study of behavior* (pp. 361–374). Dordrecht, the Netherlands: Kluwer Academic Publishers.

Pellis, S. M. (1996). Righting and the modular organization of motor programs. In K.-P. Ossenkopp, M. Kavaliers & P. R. Sanberg (Eds.), *Measuring movement and locomotion: From invertebrates to humans* (pp. 115–133). Austin, TX: Landes Company.

Pellis, S. M. (1997). Targets and tactics: The analysis of moment-to-moment decision making in animal combat. *Aggressive Behavior*, 23, 107–129.

Pellis, S. M. (2011). Head and foot coordination in head scratching and food manipulation by purple swamp hens (*Porphyrio porphyrio*): Rules for minimizing the computational costs of combining movements from multiple parts of the body. *International Journal of Comparative Psychology*, 24, 255–271.

Pellis, S. M., & Bell, H. C. (2011). Closing the circle between perceptions and behavior: A cybernetic view of behavior and its consequences for studying motivation and development. *Developmental Cognitive Neuroscience*, 1, 404–413.

Pellis, S. M., & Bell, H. C. (2020). Unraveling the dynamics of dyadic interactions: Perceptual control in animal contests. In W. Mansell (Ed.), *The interdisciplinary handbook of perceptual control theory: Living control systems IV* (pp. 75–97). Oxford: Elsevier.

Pellis, S. M., & Iwaniuk, A. N. (2004). Evolving a playful brain: A levels of control approach. *International Journal of Comparative Psychology*, 17, 90–116.

Pellis, S. M., & McKenna, M. M. (1992). Intrinsic and extrinsic influences on play fighting in rats: Effects of dominance, partner's playfulness, temperament and neonatal exposure to testosterone propionate. *Behavioural Brain Research*, 50, 135–145.

Pellis, S. M., & Officer, R. C. E. (1987). An analysis of some predatory behaviour patterns in four species of carnivorous marsupials (Dasyuridae), with comparative notes on the eutherian cat *Felis catus*. *Ethology*, 75, 177–196.

Pellis, S. M., & Pellis, V. C. (1982). Do post-hatching factors limit clutch size in the Cape Barren goose (*Cereopsis novaehollandiae* Latham)? *Australian Wildlife Research*, 9, 145–149.

Pellis, S. M., & Pellis, V. C. (1987). Play-fighting differs from serious fighting in both target of attack and tactics of fighting in the laboratory rat *Rattus norvegicus*. *Aggressive Behavior*, 13, 227–242.

Pellis, S. M., & Pellis, V. C. (1988a). Play-fighting in the Syrian golden hamster *Mesocricetus auratus* Waterhouse, and its relationship to serious fighting during post-weaning development. *Developmental Psychobiology*, 21, 323–337.

Pellis, S. M., & Pellis, V. C. (1988b). Identification of the possible origin of the body target which differentiates play-fighting from serious fighting in Syrian golden hamsters *Mesocricetus auratus*. *Aggressive Behavior*, 14, 437–449.

Pellis, S. M., & Pellis, V. C. (1989). Targets of attack and defense in the play fighting by the Djungarian hamster *Phodopus campbelli*: Links to fighting and sex. *Aggressive Behavior*, 15, 217–234.

Pellis, S. M., & Pellis, V. C. (1992). Analysis of the targets and tactics of conspecific attack and predatory attack in northern grasshopper mice (*Onychomys leucogaster*). *Aggressive Behavior*, 18, 301–316.

Pellis, S. M., & Pellis, V. C. (1994). The development of righting when falling from a bipedal standing posture: Evidence for the dissociation of dynamic and static righting reflexes in rats. *Physiology & Behavior*, 56, 659–663.

Pellis, S. M., & Pellis, V. C. (1998). The play fighting of rats in comparative perspective: A schema for neurobehavioral analyses. *Neuroscience & Biobehavioral Reviews*, 23, 87–101.

Pellis, S. M., & Pellis, V. C. (2009). *The playful brain. Venturing to the limits of neuroscience*. Oxford: Oneworld Press.

Pellis, S. M., & Pellis, V. C. (2011). To whom the play signal is directed: A study of headshaking in black-handed spider monkeys (*Ateles geoffroyi*). *Journal of Comparative Psychology*, 125, 1–10.

Pellis, S. M., & Pellis, V. C. (2012). Anatomy is important, but need not be destiny: Novel uses of the thumb in aye-ayes compared to other lemurs. *Behavioural Brain Research*, 231, 378–285.

Pellis, S. M., & Pellis, V. C. (2015). Are agonistic behavior patterns signals or combat tactics – or does it matter? Targets as organizing principles of fighting. *Physiology & Behavior*, 146, 73–78.

Pellis, S. M., & Pellis, V. C. (2016). Play fighting in Visayan warty pigs (*Sus cebifrons*): Insights on restraint and reciprocity in the maintenance of play. *Behaviour*, 153, 727–747.

Pellis, S. M., & Pellis, V. C. (2017). What is play fighting and what is it good for? *Learning & Behavior*, 45, 355–366.

Pellis, S. M., & Pellis, V. C. (2018). "I am going to groom you": Multiple forms of play fighting in gray mouse lemurs (*Microcebus murinus*). *Journal of Comparative Psychology*, 132, 6–15.

Pellis, S. M., Field, E. F., & Whishaw, I. Q. (1999). The development of a sex-differentiated defensive motor-pattern in rats: A possible role for juvenile experience. *Developmental Psychobiology*, 35, 156–164.

Pellis, S. M., Gray, D., & Cade, W. H. (2009). The judder of the cricket: The variance underlying the invariance in behavior. *International Journal of Comparative Psychology*, 22, 188–205.

Pellis, S. M., Pellis, V. C., & Dewsbury, D. A. (1989). Different levels of complexity in the playfighting by muroid rodents appear to result from different levels of intensity of attack and defense. *Aggressive Behavior*, 15, 297–310.

Pellis, S. M., Pellis, V. C., & Foroud, A. (2005). Play fighting: Aggression, affiliation and the development of nuanced social skills. In R. Tremblay,

W. W. Hartup & J. Archer (Eds.), *Developmental origins of aggression* (pp. 47–62). New York: Guilford Press.

Pellis, S. M., Pellis, V. C., & Iwaniuk, A. N. (2014). Pattern in behavior: The characterization, origins and evolution of behavior patterns. *Advances in the Study of Behavior*, 46, 127–189.

Pellis, S. M., Pellis, V. C., & McKenna, M. M. (1994). A feminine dimension in the play fighting of rats (*Rattus norvegicus*) and its defeminization neonatally by androgens. *Journal of Comparative Psychology*, 108, 68–73.

Pellis, S. M., Pellis, V. C., & Nelson, J. E. (1992). The development of righting reflexes in the pouch young of the marsupial *Dasyurus hallucatus*. *Developmental Psychobiology*, 25, 105–125.

Pellis, S. M., Pellis, V. C., & Teitelbaum, P. (1991). Air-righting without the cervical righting reflex in adult rats. *Behavioural Brain Research*, 45, 185–188.

Pellis, S. M., Pellis, V. C., & Whishaw, I. Q. (1992). The role of the cortex in play fighting by rats: Developmental and evolutionary implications. *Brain, Behavior & Evolution*, 39, 270–284.

Pellis, S. M., Williams, L., & Pellis, V. C. (2017). Adult-juvenile play fighting in rats: Insight into the experiences that facilitate the development of socio-cognitive skills. *International Journal of Comparative Psychology*, 30, 1–13.

Pellis, S. M., Burke, C. J., Kisko, T. M., & Euston, D. R. (2018). 50-kHz vocalizations, play and the development of social competence. In S. Brudzynski (Ed.), *Handbook of behavioral neuroscience*, Volume 25, *Handbook of ultrasonic vocalization. A window in the emotional brain* (pp. 117–126). New York: Academic Press.

Pellis, S. M., Field, E. F., Smith, L. K., & Pellis, V. C. (1997). Multiple differences in the play fighting of male and female rats. Implications for the causes and functions of play. *Neuroscience & Biobehavioral Reviews*, 21, 105–120.

Pellis, S. M., Pellis, V. C., Manning, C. J., & Dewsbury, D. A. (1992), Supine defense in intraspecific fighting of male house mice *Mus domesticus*. *Aggressive Behavior*, 18, 373–379.

Pellis, S. M., Pellis, V. C., Pierce, J. D., Jr., & Dewsbury, D A. (1992). Disentangling the contribution of the attacker from that of the defender in the differences in the intraspecific fighting of two species of voles. *Aggressive Behavior*, 18, 425–435.

Pellis, S. M., Pellis, V. C., Chen, Y.-C., Barzci, S., & Teitelbaum, P. (1989). Recovery from axial apraxia in the lateral hypothalamic labyrinthecto-mized rat reveals three elements of contact-righting: Cephalic domin-ance, axial rotation, and distal limb action. *Behavioural Brain Research*, 35, 241–251.

Pellis, S. M., Pellis, V. C., Chesire, R. M., Rowland, N. E., & Teitelbaum, P. (1987). Abnormal gait sequence in the locomotion released by atropine in catecholamine deficient akinetic rats. *Proceedings of the National Academy of Science* (USA), 84, 8750–8753.

Pellis, S. M., Blundell, M. A., Bell, H. C., Pellis, V. C., Krakauer, A. H., & Patricelli, G. L. (2013). Drawn into the vortex: The facing-past encounter and combat in lekking male greater sage-grouse (*Centrocercus urophasianus*). *Behaviour*, 150, 1567–1599.

Pellis, S. M., McKenna, M. M., Field, E. F., Pellis, V. C., Prusky, G. T., & Whishaw, I. Q. (1996). Uses of vision by rats in play fighting and other close quarter social interactions. *Physiology & Behavior*, 59, 905–913.

Pellis, S. M., Pellis, V. C., Himmler, B. T., Modlińska, K., Stryjek, R., Kolb, B., & Pisula, W. (2019). Domestication and the role of social play on the development of socio-cognitive skills in rats. *International Journal of Comparative Psychology*, 32, 1–12.

Pellis, V. C., Pellis, S. M., & Teitelbaum, P. (1991). A descriptive analysis of the post-natal development of contact-righting in rats (*Rattus norvegicus*). *Developmental Psychobiology*, 24, 237–263.

Penfield, W., & Boldrey, E. (1937). Somatic motor and sensory representation in the cerebral cortex of man as studied by electrical stimulation. *Brain*, 60, 389–443.

Penfield, W., & Rasmussen, T. (1950). *The cerebral cortex of man. A clinical study of localization of function*. New York: Macmillan.

Petri, H. L., & Govern, J. M. (2004). *Motivation. Theory, research, and applications*. Belmont, CA: Wadsworth.

Pfeifer, R., & Bongard, J. (2007). *How the body shapes the way we think*. Cambridge, MA: MIT Press.

Pierce, J. D., Jr., Pellis, V. C., Dewsbury, D. A., & Pellis, S. M. (1991). Targets and tactics of agonistic and precopulatory behavior in montane and prairie voles: Their relationship to juvenile play fighting. *Aggressive Behavior*, 17, 337–349.

Poletaeva, I., & Zorina, Z. (2015). Extrapolation ability in animals and its possible links to exploration, anxiety, and novelty seeking. In M. Nadin (Ed.), *Learning from the past* (pp. 415–430). Berlin, Germany: Springer.

Powers, W. T. (2005). *Behavior: The control of perception*, 2nd edition. New Canaan, CN: Benchmark Publications.

Powers, W. T. (2009). *Living control systems: The fact of control*. New Canaan, CN: Benchmark Publications.

Railsback, S. F., & Grimm, V. (2011). *Agent-based and individual based modeling: A practical guide*. Princeton: Princeton University Press.

Reinhart, C. J., McIntyre, D. C., & Pellis, S. M. (2004). The development of play fighting in kindling-prone (FAST) and kindling-resistant (SLOW) rats: How does the retention of phenotypic juvenility affect the complexity of play? *Developmental Psychobiology*, 45, 83–92.

Reinhart, C. J., McIntyre, D. C., Metz, G. A., & Pellis, S. M. (2006). Play fighting between kindling-prone (fast) and kindling-resistant (slow) rats. *Journal of Comparative Psychology*, 120, 19–30.

Roberto, M. E., & Brumley, M. R. (2014). Prematurely delivered rats show improved motor coordination during sensory-evoked motor responses compared to age-matched controls. *Physiology & Behavior*, 130, 75–84.

Ronca, A. E., & Alberts, J. R. (2000). Effects of prenatal spaceflight on vestibular responses of neonatal rats. *Journal of Applied Physiology*, 89, 2318–2324.

Rychlowska, M., Cañadas, E., Wood, A., Krumhuber, E. G., Fischer, A., & Niedenthal, P. M. (2014). Blocking mimicry makes true and false smiles look the same. *PLoS ONE*, 9(3), e90876. doi:10.1371/journal.pone.0090876

Sandhu, S., Gulrez, T., & Mansell, W. (2020). Behavioral anatomy of a hunt: Using dynamic real-world paradigm and computer vision to compare human user-generated strategies with prey movement varying in predictability. *Attention, Perception & Psychophysics*, 82, 3112–3123

Schank, J. C., & Alberts, J. R. (1997). Self-organized huddles of rat pups modeled by simple rules of individual behavior. *Journal of Theoretical Biology*, 189, 11–25.

Schank, J. C., May, C. J., Tran, J. T., & Joshi, S. S. (2004). A biorobotic investigation of Norway rat pups (*Rattus norvegicus*) in an arena. *Adaptive Behavior*, 12, 161–173.

Sherrington, C. S. (1906). *The integrative action of the nervous system.* New York: Scribner's.

Sherwood, L., & Ward, C. (2018). *Human physiology: From cells to systems.* Toronto, ON: Nelson Education.

Siegel, H. I. (1985). Male sexual behavior. In H. I. Siegel (Ed.), *The hamster. Reproduction and behavior* (pp. 191–206). New York: Plenum Press.

Silverman, P. (1978). *Animal behaviour in the laboratory.* New York: Pica Press.

Sinnamon, H. M. (1993). Preoptic and hypothalamic neurons and the initiation of locomotion in the anesthetized rat. *Progress in Neurobiology*, 141, 323–44.

Siviy, S. M. (2016). A brain motivated to play: Insights into the neurobiology of playfulness. *Behaviour*, 153, 819–844.

Siviy, S. M., & Panksepp, J. (1987). Sensory modulation of juvenile play in rats. *Developmental Psychobiology*, 20, 39–55.

Siviy, S. M., Baliko, C. N., & Bowers, K. S. (1997). Rough-and-tumble play behavior in Fischer-344 and Buffalo rats: Effects of social isolation. *Physiology & Behavior*, 61, 597–602.

Siviy, S. M., Crawford, C. A., Akopian, G., & Walsh, J. P. (2011). Dysfunctional play and dopamine physiology in the Fisher 344 rat. *Behavioural Brain Research*, 220, 294–304.

Siviy, S. M., Love, N. J., DeCicco, B. M., Giordano, S. B., & Seifert, T. L. (2003). The relative playfulness of juvenile Lewis and Fischer-344 rats. *Physiolology & Behavior*, 80, 385–394.

Skinner, B. F. (1938). *The behavior of organisms*. New York: Appleton Century Crofts.

Smith, P. K. (1997). Play fighting and real fighting. Perspectives on their relationship. In A. Schmitt, K. Atzwanger, K. Grammar, & K. Schäfer (Eds.), *New aspects of human ethology* (pp. 47–64). New York: Plenum Press.

Staddon, J. E. R. (2016). *Adaptive behavior and learning*. 2nd edition. Cambridge: Cambridge University Press.

Stark, R. A., Harker, A., Salamanca, S., Pellis, S. M., Li, F., & Gibb, R. L. (2020). Development of ultrasonic calls in rat pups follows similar patterns regardless of isolation distress. *Developmental Psychobiology*, 62, 617–630.

Strack, F., Martin, L. L., & Stepper, S. (1988). Inhibiting and facilitating conditions of the human smile: A nonobtrusive test of the facial feedback hypothesis. *Journal of Personality & Social Psychology*, 54, 768–777.

Suomi, S. J. (2005). Genetic and environmental factors influencing the expression of impulsive aggression and serotonergic functioning in rhesus monkeys. In R. E. Tremblay, W. W. Hartup & J. Archer (Eds.), *Developmental origins of aggression* (pp. 63–82). New York: Guilford Press.

Szechtman, H., Ornstein, K., Teitelbaum, P., & Golani, I. (1985). The morphogenesis of stereotyped behavior induced by the dopamine receptor agonist apomorphine in the laboratory rat. *Neuroscience*, 14, 783–798.

Taylor, G. T. (1980). Fighting in juvenile rats and the ontogeny of agonistic behavior. *Journal of Comparative & Physiological Psychology*, 94, 953–961.

Teitelbaum, P. (1982). Disconnection and antagonistic interaction of movement subsystems in motivated behavior. In A. R. Morrison & A. L. Strick (Eds.), *Changing concepts of the nervous system: Proceedings of the first institute of neurological sciences symposium in neurobiology* (pp. 467–487). New York: Academic Press.

Teitelbaum, P. (2012). Some useful insights for graduate students beginning their research in physiological psychology: Anecdotes and attitudes. *Behavioural Brain Research*, 231, 234–249.

Teitelbaum, P., Cheng, M.-F., & Rozin, P. (1969). Development of feeding parallels its recovery after hypothalamic damage. *Journal of Comparative & Physiological Psychology*, 67, 430–441.

Teitelbaum, P., Schallert, T., & Whishaw, I. Q. (1983). Sources of spontaneity in motivated behavior. In E. Satinoff & P. Teitelbaum (Eds.), *Motivation* (pp. 23–66). Boston, MA: Springer.

Teitelbaum, P., Wolgin, D. L., De Ryck, M., & Marin, O. S. (1976). Bandage-backfall reaction: Occurs in infancy, hypothalamic damage, and catalepsy. *Proceedings of the National Academy of Science*, USA, 73, 3311–3314.

Thelen, E. (1995). Motor development. A new synthesis. *American Psychologist*, 50, 79–95.

Thor, D. H., & Flannelly, K. J. (1978). Sex-eliciting behavior of the female rat: Discrimination of receptivity by anosmic and intact males. *Behavioral Biology*, 23, 326–340.

Thor, D. H., & Holloway, W. R., Jr. (1986). Social play soliciting by male and female juvenile rats: Effects of neonatal androgenization and sex of cagemates. *Behavioral Neuroscience*, 100, 275–279.

Tilney, F. (1933). Behavior in its relation to the development of the brain: Part II. Correlation between the development of the brain and behavior in the albino rat from embryonic states to maturity. *Bulletin of the Neurology Institute of New York*, 3, 252–258.

Timberlake, W. (2001). Motivational modes in behavior systems. In R. R. Mowrer & S. B. Klein (Eds.), *Handbook of contemporary learning theories* (pp. 155–210). San Francisco, CA: Erlbaum.

Tinbergen, N. (1951). *The study of instinct*. Oxford, UK: Clarendon Press.

Tinbergen, N. (1963). On aims and methods of ethology. *Zeitschrift für Tierpsycholgie*, 20, 410–433.

Tinbergen, N., & Lorenz, K. (1938). Taxis und Instinkthandlung in der Eirollbewegung der Graugans. *Zeitschrift für Tierpsycholgie*, 2, 1–29.

Tomkiewicz, S. M., Fuller, M. R., Kie, J. G. & Bates, K. K. (2010). Global positioning system and associated technologies in animal behaviour and ecological research. *Philosophical Transactions of the Royal Society* B, 365, 2163–2176.

Troiani, D., Petrosini, L., & Passani, F. (1981). Trigeminal contribution to the head righting reflex. *Physiology & Behavior*, 37, 157–160.

Turner, J. S. (2007). *The tinkerer's accomplice. How design emerges from life itself.* Cambridge, MA: Harvard University Press.

Vanderschuren, L. J. M. J., Achterberg, E. J. M., & Trezza, V. (2016). The neurobiology of social play and its rewarding value in rats. *Neuroscience & Biobehavioral Reviews, 70,* 86–105.

Vasey, P. L., Foroud, A., Duckworth, N., & Kovacovsky, S. D. (2006). Male-female and female–female mounting in Japanese macaques: A comparative study of posture and movement. *Archives of Sexual Behavior, 35,* 116–128.

Vergara-Aragon, P., Gonzalez, C. L. R., & Whishaw, I. Q. (2003). A novel skilled-reaching impairment in paw supination on the "good" side of the hemi-Parkinson rat improves with rehabilitation. *The Journal of Neuroscience, 23,* 579–586.

Von Holst, E. (1973). *The behavioral physiology of animals and man* (trans. Martin, R.). Coral Gables, FL: University of Miami Press.

Von Uexküll, J. (1934/2010). A foray into the worlds of animals and humans. Translated by J. D. O'Neil and reprinted by the University of Minnesota Press: Minneapolis, MN.

Vorhees, C. V., & Williams, M. T. (2006). Morris water maze: Procedures for assessing spatial and related forms of learning and memory. *Nature Protocols, 1,* 848–858.

Wainwright, P. C., Mehta, R. S., & Higham, T. E. (2008). Stereotypy, flexibility and coordination: Key concepts in behavioral functional morphology. *Journal of Experimental Biology, 211,* 3523–3528.

Wallace, D. G., Choudhry, S., & Martin, M. N. (2006). Comparative analysis of movement characteristics during dead-reckoning-based navigation in humans and rats. *Journal of Comparative Psychology, 120,* 331–344.

Wallace, D. G., Hines, D. J., Pellis, S. M., & Whishaw, I. Q. (2002). Vestibular information is required for dead reckoning in the rat. *Journal of Neuroscience, 22,* 10009–10017.

Walther, F. R. (1984). *Communication and expression in hoofed mammals.* Bloomington: Indiana University Press.

Walton, K. D., Harding, S., Anschel, D., Harris, Y. T., & Llinás, R. (2005). The effects of microgravity on the development of surface righting in rats. *Journal of Physiology, 565,* 593–608.

Waterman, J. M. (2010). The adaptive function of masturbation in a promiscuous African ground squirrel. *PLoS ONE, 5*(9), e13060. doi:10.1371/journal.pone.0013060

Webb, B. (2001). Can robots make good models of biological behavior? *Behavioral & Brain Sciences, 24,* 1033–1050.

Webb, B., & Consi, T. R. (2001). *Biorobotics: Methods and applications*. Menlo Park, CA: American Association for Artificial Intelligence.

Whishaw, I. Q. (1988). Food wrenching and dodging: Use of action patterns for the analysis of sensorimotor and social behavior in the rat. *Journal of Neuroscience Methods*, 24, 169–178.

Whishaw, I. Q., & Gorny, B. (1994a). Arpeggio and fractionated digit movements used in prehension by rats. *Behavioural Brain Research*, 60, 15–24.

Whishaw, I. Q., & Gorny, B. (1994b). Food wrenching and dodging: Eating time estimates influence dodge probability and amplitude. *Aggressive Behavior*, 20, 35–47.

Whishaw, I. Q., & Kolb, B. (1985). The mating movements of male decorticate rats: Evidence for subcortically generated movements by the male but regulation of approaches by the female. *Behavioural Brain Research*, 17, 171–191.

Whishaw, I. Q., & Miklyaeva, E. I. (1996). A rat's reach should exceed its grasp: Analysis of independent limb and digit use in the laboratory rat. In P.-K. Ossenkopp, M. Kavaliers & P. R. Sanburg (Eds.), *Measuring movement and locomotion: From invertebrates to humans* (pp. 135–169). Austin, TX: Landes.

Whishaw, I. Q., & Pellis, S. M. (1990). The structure of skilled forelimb reaching in the rat: A proximally driven stereotyped movement with a single rotatory component. *Behavioural Brain Research*, 41, 49–59.

Whishaw, I. Q., Cassel, J.-C., & Jarrard, L. E. (1995). Rats with fimbria-fornix lesions display a place response in a swimming pool: A dissociation between getting there and knowing where. *The Journal of Neuroscience*, 15, 5779–5788.

Whishaw, I. Q., Dringenberg, H. C., & Pellis, S. M. (1992). Forelimb use in free feeding by rats: Motor cortex aids limb and digit positioning. *Behavioural Brain Research*, 48: 113–125.

Whishaw, I. Q., O'Connor, W. T., & Dunnett, S. B. (1986). The contributions of motor cortex, nigrostriatal dopamine and caudate-putamen to skilled forelimb use in the rat. *Brain*, 109, 805–843.

Whishaw, I. Q., Pellis, S. M., & Gorny, B. P. (1992a). Medial frontal cortex lesions impair the aiming component of rat reaching. *Behavioural Brain Research*, 50, 93–104.

Whishaw, I. Q., Pellis, S. M., & Gorny, B. P. (1992b). Skilled reaching in rats and humans: Evidence for parallel development or homology. *Behavioural Brain Research*, 47, 59–70.

Whishaw, I. Q., Pellis, S. M., & Pellis, V. C. (1992). A behavioral study of the contributions of cells and fibers of passage in the red nucleus of the rat to postural righting, skilled movements, and learning. *Behavioural Brain Research*, 52, 29–44.

Whishaw, I. Q., Sarna, J., & Pellis, S. M. (1998). Evidence for rodent-common and species-typical limb and digit use in eating derived from a comparative analysis of ten rodent species. *Behavioural Brain Research*, 96, 79–91.

Whishaw, I. Q., Whishaw, P., & Gorny, B. (2008). The structure of skilled forelimb reaching in the rat: A movement rating scale. *Journal of Visual Experiments*, 18, e816. doi:10.3791/816.

Whishaw, I. Q., Pellis, S. M., Gorny, B. P., & Pellis, V. C. (1991). The impairments in reaching and the movements of compensation in rats with motor cortex lesions: A videorecording and movement notation analysis. *Behavioural Brain Research*, 42, 77–91.

Whishaw, I. Q., Suchowersky, O., Davis, L., Sarna, J., Metz, G. A., & Pellis, S. M. (2002). A qualitative analysis of reaching-to-grasp movements in human Parkinson's disease (PD) reveals impairments in coordination and rotational movements of pronation and supination: A comparison to deficits in animal models of PD. *Behavioural Brain Research*, 133, 165–176.

Wiley, R. H. (1973). The strut display of male sage grouse – a 'fixed' action pattern. *Behaviour*, 47, 129–152.

Wilmer, A. H. (1991). Behavioral deficiencies of aggressive 8–9-year-old boys: An observational study. *Aggressive Behavior*, 17, 135–154.

Windle, W. F., & Fish, M. W. (1932). The development of the vestibular righting reflex in the cat. *Journal of Comparative Neurology*, 54, 85–96.

Wright, J. M., Gourdon, J. M., & Clarke, P. B. (2010). Identification of multiple call categories within the rich repertoire of adult rat 50-kHz ultrasonic vocalizations: Effects of amphetamine and social context. *Psychopharmacology*, 211, 1–13.

Index